New Prospects of Maize

Edited by Prashant Kaushik

Published in London, United Kingdom

New Prospects of Maize
http://dx.doi.org/10.5772/intechopen.107793
Edited by Prashant Kaushik

Contributors
Marcos Arturo Ferreira Agüero, Felisa de Los Ríos Peixoto, Arsenio Benítez Sánchez, José Augusto Velásquez, Gustavo Daniel Vega-Brítez, Nelson David Lesmo Duarte, Matheus Francisco Acosta Resquín, Rupinder Chandel, Mahesh Kumar Narang, Surinder Singh Thakur, Cyndy Margarita Garcia-Weyandt, Bharat Kumar Pradhan, Ghanashyam Sharma, Maryam Batool, Elaissaoui Abdellah, El Khalfi Chaima, Harkani Assia, Ouhemi Hanane, Benabdelouahab Tarik, Smruti Ranjan Padhan, Pratikshya Mishra, Sibananda Darjee, Sushmita Saini, Shankar Lal Jat, Radheshyam, Shivani Ranjan, Soumya Ranjan Padhan, María Berenice González-Maldonado, Miguel Mauricio Correa-Ramírez, Mónica Yazmín Flores-Villegas, Paul L. Woomer, Welissa Mawia Mulei, Dries Roobroeck, Vikas Kumar Singh, Saksham Pundir, Deepti Chaturvedi, Arshdeep Kaur, Abhishek Pandey, Shreya Mandal, Rahul Kumar, Harish Pal Bhati, Ramesh Singh, Chittaranjan Kole, Prashant Kaushik, Parmdeep Singh Dhanda, Alpa Yadav

First published in London, United Kingdom, 2024 by IntechOpen
IntechOpen is the global imprint of INTECHOPEN LIMITED, registered in England and Wales, registration number: 11086078, 5 Princes Gate Court, London, SW7 2QJ, United Kingdom
Printed in Croatia

British Library Cataloguing-in-Publication Data
A catalogue record for this book is available from the British Library

Additional hard and PDF copies can be obtained from orders@intechopen.com

New Prospects of Maize
Edited by Prashant Kaushik
p. cm.

This title is part of the Agricultural Sciences Book Series, Volume 2
Topic: Agronomy and Horticulture
Series Editor: W. James Grichar
Topic Editor: Ibrahim Kahramanoglu
Associate Topic Editors: Murat Helvaci and Olga Panfilova

Print ISBN 978-1-83768-631-5
Online ISBN 978-1-83768-632-2
eBook (PDF) ISBN 978-1-83768-633-9
ISSN 3029-052X

For EU product safety concerns:
IN TECH d.o.o., Prolaz Marije Krucifikse Kozulić 3, 51000 Rijeka, Croatia,
info@intechopen.com or visit our website at intechopen.com.

IntechOpen Book Series

Agricultural Sciences

Volume 2

Aims and Scope of the Series

The importance of agriculture cannot be overstated. It helps sustain life, as it gives us the food we need to survive and provides opportunities for economic well-being. Agriculture helps people prosper around the world and combines the creativity, imagination, and skill involved in planting crops and raising animals with modern production methods and new technologies. This series includes two main topics: Agronomy and Horticulture, and Animal Farming. This series will help readers better understand the intricacies of production agriculture and provide the new knowledge that is required to be successful. The success of a farmer in modern agriculture requires knowledge of events happening locally as well as globally that impact input decisions and ultimately determine net profit.

Meet the Series Editor

W. James Grichar has been employed with Texas A&M AgriLife Research for over 45 years with an emphasis on research in agronomy, plant pathology, and weed science. He obtained his BS from Texas A&M in 1972 and his Masters of Plant Protection in 1975. He has published 195 journal articles, over 330 research reports and briefs, 11 book chapters, and over 300 abstracts of profession meetings. He also directs research in many crops including corn, grain sorghum, peanuts, and sesame. He has held various positions in different professional societies including the American Peanut Research and Education Society, Southern Weed Science Society, and Texas Plant Protection Conference in addition to being Associate Editor for Peanut Science and Weed Technology. Significant accomplishments have included spearheading efforts to determine the optimum planting time for soybean production along the upper Texas Gulf Coast. These efforts have shown growers that soybean yields can be improved by 10 to 20% by following a late March to early April plant date. He also has been instrumental in developing a herbicide program for peanut production in the south Texas growing region. Through the development and use of herbicides that are effective against major weed problems in the south Texas region, peanut yields have increased by 25 to 30%.

Meet the Volume Editor

Dr. Prashant Kaushik is a venerated genomics, molecular biology, and bioinformatics expert, boasting a decade of exceptional academic and research contributions. His scholarly output includes more than 250 peer-reviewed articles and 10 books, reflecting his deep engagement with cutting-edge methodologies such as RNA sequencing and computational biology. Dr. Kaushik is renowned for his dedication to science and effective communication, fostering significant collaborations between academic institutions and the agricultural industry. His influential research and leadership have made him a key innovator in agricultural sciences, and his commitment is shaping the future of sustainable crop development.

Contents

Preface

In the ambit of agricultural innovation and crop science, maize (*Zea mays*) epitomizes the pinnacle of human agricultural development and the evolutionary success facilitated by scientific intervention. *New Prospects of Maize* is an edited compendium that collates pioneering research and comprehensive reviews, offering a fresh perspective on this globally pivotal cereal crop. This preface provides a concise introduction to the breadth of topics and the consortium of expertise encapsulated within this volume.

The odyssey into the world of maize begins with an exposition on its vast genetic and cultivar diversity in "Exploring the Diversity of Maize." This section takes the reader through the intricate maze of maize's genetic endowment, which has been expansively modified through both natural and anthropogenic selection, revealing the potential locked within its genome.

Advancing to the agronomic aspects, "Nutrient Planning and Control of Maize" probes into the agronomic practices that determine the nutrient dynamics of maize cultivation. It offers a synthesis of current practices and emerging innovations aimed at enhancing the nutrient use efficiency of maize, which is crucial for optimizing yield and environmental sustainability.

The volume then delves into the realms of agroeconomics and agricultural engineering with "Scaling Mechanization and Profitability in Maize Cultivation". Herein, the transformative impact of mechanization on the economics of maize production is scrutinized, with a focus on how technological advancements can be harmonized with agricultural practices to bolster productivity and economic viability.

In "Maize Breeding and Genomics", the narrative shifts to the molecular vista, charting the recent progressions in maize breeding accelerated by genomics and biotechnological interventions. The discourse encapsulates the latest methodologies and genomic tools that are reshaping the landscape of maize breeding, aiming at the development of varieties with enhanced traits such as yield, nutritional quality, and stress tolerance.

Concluding the volume is "Modern Ways to Insect Management", which confronts the perpetual battle against insect pests in maize cultivation. This section elucidates contemporary strategies in pest management, emphasizing integrated approaches that balance effectiveness with ecological and environmental considerations.

Gratitude is owed to all those who contributed to the editorial journey, whose meticulous efforts and scholarly insights have honed the precision and depth of the volume.

In adherence to the highest scientific and ethical standards, this preface and the volume it precedes have been composed with a profound respect for the scientific endeavor and its contribution to global food security. It is with profound optimism that *New Prospects of Maize* is presented to the scientific community, in the hope that it will serve as a catalyst for further scientific exploration and innovation in maize research.

Dr. Prashant Kaushik
Chaudhary Charan Singh Haryana Agricultural University,
Hisar, India

Section 1

Exploring the Diversity of Maize

Chapter 1

Exploring the Diversity of Maize (*Zea mays* L.) in the Khangchendzonga Landscapes of the Eastern Himalaya

Ghanashyam Sharma and Bharat Kumar Pradhan

Abstract

The Sikkim Himalaya is a distinguished hub of maize biodiversity, housing a wide range of genetic resources cultivated at altitudes from 300 to 2500 m elevations. From 2010 to 2022, a field investigation combined traditional knowledge and scientific methods to morphologically characterize maize, supplemented by relevant literature. The objective was to evaluate indigenous maize varieties in the region since the 1960s. The research classified maize landraces into four groups: primitive landraces, preserved traditional popcorn races; advanced or derived landraces, selectively bred for desirable traits; recent introductions from other regions; and hybrid maize varieties resulting from crossbreeding. About 31 maize landraces were listed, emphasizing the urgent need for in-depth genetic characterization. Notably, *Murali Makai, Seti Makai, Pahenli Makai, Rato Makai, Baiguney Makai, Gadbadey Makai, Tempo-Rinzing*, and *Lachung Makai* adapted well to altitudes of 300–2500 m, showing variations in agronomic and quality traits, as well as resistance to environmental stresses. Primitive maize cultivars in the Northeastern Himalayas of India have generated interest among researchers for their high prolificacy and their link to the origin and evolution of maize. Prioritization at the species level and within specific geographic regions is necessary due to the dynamic demand for germplasm. Conservation of certain maize germplasm is crucial for food security, livelihoods, climate resilience, and research. The study identified potential risks of germplasm extinction or erosion, emphasizing the need for urgent actions to safeguard these genetic resources.

Keywords: Khangchendzonga landscapes, maize landraces, quality traits, genetic diversity, eastern Himalaya

1. Introduction

Maize (*Zea mays* L.) is a significant cereal crop worldwide, widely cultivated and highly valued for its diverse uses. In high-income countries, it serves as a primary source of feed and industrial products, while in sub-Saharan Africa (SSA), Asia, and Latin America, it plays a crucial role in providing food, feed, and nutritional security for some of the world's poorest regions [1]. Originating from Mexico, maize has extended its presence to critical agroecologies in tropical, subtropical, and temperate regions across the globe.

The center of origin for maize has been established as the Mesoamerican region, encompassing present-day Mexico and Central America [2]. Maize cultivation was introduced to the rest of the world during the sixteenth century [3, 4], and in India, it was brought by the Portuguese in the seventeenth century [5–7]. In India, there is speculation that maize was initially introduced in the North-West Himalayas by traders, from where it subsequently spread to the Himalayan region [8]. This includes the introduction of primitive forms of maize in the old world, including the Himalayas during the pre-Columbian period [9].

The exceptional genetic diversity of maize is evident from well-defined landraces found in various countries. Landraces, which serve as a valuable reservoir of useful genes and alleles, are the germplasm preserved by farmers over extended periods. These landraces were developed and selected to thrive under specific environmental conditions and fulfill local food preferences.

India is a significant centre of diversity for maize landraces, with extensive variations observed in plant, ear, and tassel characteristics in the northeastern and northwestern highlands. However, relatively less varietal diversity is found in the plains of India [10]. Anderson [11] was particularly impressed with the diversity of maize in the north-eastern Himalayan region and believed that maize had an ancient Asiatic origin. Dhawan [12] renamed a productive popcorn variety from Sikkim as 'Sikkim Primitive' due to its distinctive characteristics. This landrace exhibits unique physiological features, such as the absence of apical dominance, prolificacy (5–9 ears), uniformity in ear size, erect leaves, a top-bearing habit, and drooping tassels for efficient fertilization [13]. The Sikkim Primitive maize is commonly used as fodder for livestock by local farmers. The primitive maize strain, also known as the SP strain, comprises 13 different strains distributed throughout the NEH region, excluding Sikkim [14].

A significant portion of the genetic diversity in maize remains unexplored, partly due to the challenge of identifying valuable genetic variations within local varieties [15]. This limited utilization of diverse genetic resources can be attributed to the lack of understanding of the agronomic performance and genetic makeup of these landraces. Prasanna and Sharma [10] addressed this bottleneck by conducting the first comprehensive molecular characterization and population genetic analysis of 17 Indian maize landraces sourced from diverse agroecologies. The study utilized 27 microsatellite or simple sequence repeat (SSR) markers to reveal a higher level of genetic divergence among the Northeastern Hill (NEH) landraces when compared to those from other regions.

Maize is a crucial crop in Sikkim, although its current productivity is quite low, with an average of 1750 kg/ha [16]. Given this, it is important to develop varieties that are well-suited to the region's unique conditions. To accomplish this, breeders require an understanding of character associations and path analysis.

Landraces are a type of germplasm that has been preserved by farmers over the course of decades or even centuries. They have been carefully selected to adapt to specific environmental conditions and meet local food preferences. While maize originally came from Mexico, it is now widely grown across the world, and India has a particularly rich array of maize landraces, particularly in the North-Eastern Himalayan region [10]. These landraces contain many valuable genes that could be used for allele mining and population improvement. In recent years, molecular analysis of maize landraces from the Americas and Europe has led to significant insights into their diversity, genetic structure, and global migration routes. It is essential to have a comprehensive understanding of the genetic and agronomic characteristics of

DOI: http://dx.doi.org/10.5772/intechopen.112566

landraces for effective management and use in breeding programs. In this review, we discuss the extensive diversity of maize landraces globally and in India in Sikkim and NEH, and explore the potential for harnessing this vast genetic resource.

2. Study area

The study area is situated in the Indian part (26°29′13.56″ to 28°7′51.6″ N and 87°59′1.32″ to 89°53′42.96″ E) of the Khangchendzonga Landscape (KL) in the eastern Himalayas, which is a significant global biodiversity hotspot. KL represents a unique blend of biodiversity, bio-cultural richness, and distinctive geo-climatic features. The Indian portion of KL covers 14,061.7 sq. km and encompasses the state of Sikkim and the northern hill region of West Bengal. It stretches across a wide altitudinal range, from 40 to 8586 m (asl).

This region is crucial for sustaining the well-being of its inhabitants, as it provides diverse ecosystems and essential ecosystem services. It boasts a rich floral diversity, with over 5500 identified species, as well as recorded fauna taxa of more than 1500 species. The presence of various socio-economic and cultural diversities further enhances the significance of the region.

The study area falls under the SECURE Himalaya project, implemented by the Forest and Environment Department of the Government of Sikkim (**Figure 1**). Specifically, the project focuses on the Khangchendzonga—Upper Teesta Landscape, covering an approximate area of 4000 sq. km. Local communities hold a deep reverence for this landscape due to the legends and traditions associated with it. The landscape is characterized by its multi-ethnic composition, including Nepalese, Lepchas, Bhutias, and Tibetan Buddhists conserving rich agricultural biodiversity.

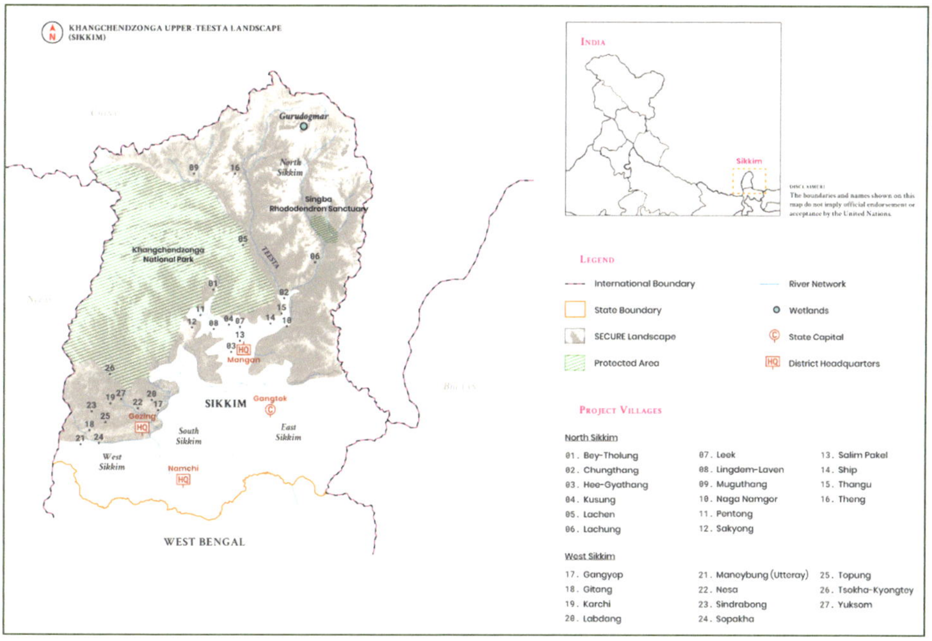

Figure 1.
Study area in the Khanchendzunga landscape based on the secure Himalaya project (https://securehimalaya.org/sikkim-landscape/).

3. Genetic diversity of maize in NEH

Singh [17] reported 10 indigenous maize varieties from the northeastern Himalayan region, namely *Nilip Mekop*, *Mikir Merakku*, *Khasi Riewhadem*, *Silken Tipang*, *Tista Mendi*, *Maidani Makka* (sub-race *Ganga*), *Cachar Gomdhan*, *Shyam Nahom*, *Asht Samsung* (sub-race *Tsungrhu*), *Mayong Sa-ah*, *Manipuri Chujak*, *Alok Sapa*, *Arun Tepi*, *Tirap NagSahypung*, and *Poorvi Botapa* (sub-race *Murli*).

As per Singh [17], *Astha Samsung*, a maize variety well adapted to high elevations above 1500 m, was commonly found in Sikkim and Nagaland. The name "Asht" refers to the presence of eight rows of kernels, a distinctive trait of this particular maize race. Samsung, on the other hand, refers to the specific location in Sikkim where this maize race was commonly cultivated. Another subrace of maize variety reported from Sikkim was *Tsungrhu*, which was also cultivated above 1500 m. The term "*Tsungrhu*" derives from the regional language of the Lotha Tribe in Nagaland, where it was extensively grown and its name signifies "maize" in their language. Similarly, *Tista Mendi* was another indigenous maize variety cultivated in the elevated regions above 1500 m along the famous River Teesta.

The races mentioned in the preceding paragraphs can be categorized into four groups for convenience: Primitive, Advanced or Derived, Recent Introductions, and Hybrid varieties. The Primitive group encompasses several races of popcorn that have differentiated at various altitudes and under diverse conditions. These races include *Poorvi Botapa*, *Murli subrace* of *Poorvi Botapa*, *Tirap NagSahypung*, *Arun Tepi*, and *Alok Sapa* [17]. The races belonging to the Primitive group display distinct traits such as popping grain morphology, prominent kernel striations, reduced kernel rows, smaller ears with higher ear numbers, tassels with fewer branches, shorter internodes, narrower leaves, and relatively higher ear placement. These races are widely distributed across the eastern Himalayan region, specifically in Assam, Nagaland, Manipur, Arunachal Pradesh, Sikkim, and Bhutan. They thrive at elevations ranging from 60 to above 2000 m, under the conditions of traditional cultivation. One fascinating aspect of the *Murli* subrace of *Poorvi Botapa* is its remarkable similarity to the reconstructed ancestral form of maize, as documented by Mangelsdorf and his collaborators [18, 19]. Notably, evidence suggests a significant differentiation in the cytoplasm between the primitive subrace of Murli and the evolved types, as evidenced by reciprocal variations in various traits observed in Sikkim and Assam [20]. The *Tirap Nagsahypung* race possesses distinctive characteristics concerning leaf size, shape, arrangement, and number. Its leaves are erect, small, numerous, and tend to cluster toward the tassel.

Singh [17] examined the second group, which included *Manipuri Chujak*, *Mayong Sa-ah*, *Asht Samsung*, *Tsungrhu* subrace of *Asht Samsung*, *Shyam Nahom*, *Cachar Gomdham*, *Mainani Makka*, and its subrace *Ganga*. These cultivars displayed distinct characteristics, such as large flinty grains exhibiting a wide array of endosperm colors, including white, cherry, red, purple, and various shades in between. They featured a smaller number of but larger ears and exhibited an earlier maturation compared to the races found in the primitive group. The leaf structure of these cultivars varied, ranging from semi-erect to flat. The collected varieties from the region exhibited limited diversity, predominantly consisting of early-flint types that closely resembled Cuban flints and northern flints. Consequently, they were classified within the *Maidani Makka* race and its subrace *Ganga*.

Singh [17] classified the third group, which includes races such as *Tista Mendi* and *Silken Tipang*. *Tista Mendi*, primarily found at higher elevations, demonstrated semi-dent grains in shades of yellow and red, accompanied by fewer but larger ears. Singh noted that *Tista Mendi* was a recent introduction that not only established itself as a distinct race but also underwent hybridization with older races, leading to the emergence of new hybrid varieties. On the other hand, *Silken Tipang* exhibited grains that exhibited popping characteristics when subjected to heat, along with a reduced number of ears. It stood out with its larger ear surfaces and more distinct grain characteristics compared to the primitive group. Singh believed that *Silken Tipang* was introduced during the early 1960s from neighboring countries like Burma and is currently confined to specific regions in Arunachal Pradesh, bordering Burma.

Singh [17] discussed the fourth group, comprising *Khasi Riewhadem*, *Mikir Merakku*, and *Nilip Mekop*. According to Singh's classification, these races were the outcome of hybridization between primitive types and advanced races. Notably, certain collections belonging to the *Tista Mendi* and *Tirap Nag-Sahypung* races exhibited remarkable resistance to the corn borer, *Chilo Zonellus*, even under natural infestation. Furthermore, varieties such as *Arun Tepi*, *Alok Sapa*, and all collections belonging to the *Tirap Nag-Sahypung* race displayed notable resistance to leaf blight caused by *Helminthosporium maydis*. Furthermore, it is worth noting that all entries from the *Manipuri Chujak* and *Mayong Sa-ah* races, along with the collections forming the *Tirap Nag-Sahypung* race, exhibited notable resistance to downy mildew, which is caused by the pathogen *Sclerospora philippinensis*. These varieties show great potential for cultivation under organic conditions, given their inherent resistance to this devastating disease.

The aforementioned discussion highlights the need for in-depth research on the genetic characterization of each variety, which remains unexplored in the region. Further investigation into the genetic traits, molecular markers, and population structure of these maize races is crucial for a comprehensive understanding of their genetic composition and diversity. Such research endeavors would contribute significantly to the conservation, utilization, and improvement of these indigenous maize varieties in the region.

4. Some basic characteristics of maize varieties

In the last 30 years, Sikkim and other Northeastern region have documented several fascinating local variations of maize that possess unique traits specific to each type of landrace. Some notable examples include *Bancharey-makai*, which is a high-altitude maize variety with yellow, flint kernels. *Badam-topo* stands out as a popcorn variety, while *Chakhou chujak* is known for its aromatic, soft, and sticky properties. *Chepti-makai* is a white, dent-type maize with distinct kernels. *Chujak* is an aromatic popcorn variant, whereas *Darikincho* is characterized by its small, yellow, hard kernels. *Fingdong* is an aromatic popcorn with a distinctive *flavor*. *Gadbade-makai* displays white kernels with occasional purple flint kernels, while *Kaali-makai* is recognized for its dark purplish-black color. *Kholakitti* is a sticky variety, while *Kuchung dari* is an orange-colored popcorn with flint kernels. *Kuchung takmar* exhibits a mix of yellow, white, purple, and red kernels with a flint texture. *Kukharey-makai* is dwarf, high-altitude maize, and *Kukidolong-makai* is a flint variety. *Lachung-makai* variety displays multiple colors and possesses tolerance to cold conditions. *Nepali Sappa* is unique with three cobs per plant, *Pahenli-makai* features yellow/orange flint kernels, and *Pahenli-makai* is a light dent type. *Phensong-makai* stands out with its cob length of up to 30 cm, while *Putali-makai* is characterized by its multi-colored appearance.

Rato-makai is a dark red maize variant, *Sathiya-makai* is an early-maturing type, and *Seti-makai* is a white, soft variety. *Tanee-makai* is a popcorn type, and *Tista Mehdi-makai* is a flint variety.

The existence of primitive maize landraces in Sikkim and other Northeastern states of India, situated in the Himalayan region, implies a possible alternative origin for this crop. The presence of a diverse collection of maize landraces in Sikkim further strengthens the notion that the Northeastern states may serve as secondary centres of origin for maize. These landraces exhibit significant morphological diversity, as documented by Sharma et al. [21]. This observation highlights the importance of studying the genetic and morphological characteristics of these landraces to unravel their evolutionary history and conservation significance.

In this region, a variety of intriguing local types and landraces with distinct traits have been documented. They include:

1. Aromatic and sticky kernels: *Fingdong* (aromatic popcorn), *Chujak* (aromatic popcorn), *Chakhou chujak* (aromatic, soft, sticky), *Kholakitthi* (sticky).

2. Popcorn varieties: *Badam topo, Tanee*.

3. Flint types: *Kukidolong*, *Kuchung dari*, *Bacherey*, *Kuchung tamar*, *Kukharey* (dwarf, suitable for high altitudes).

4. Dent kernel varieties: *Gadbade*, *Seti*, *Chepti makai* (soft opaque cap), *Pahenli* (light dent).

Apart from these, sporadic collections of early local types have also been made, such as *Ambo*, *Riewhadem* (early maturing), *Vaimin* (3 months), *Pahari makai* (adapted to mid-to-high altitudes and cold-hardy), *Nepali Sappa* (3 cobs/plant). Furthermore, modern cultivars and newly introduced landraces like *Mampokmendi, Taminlamendi, Maromendi* have been documented (**Figures 2–5**).

A germplasm exploration and collection program conducted in Sikkim has revealed the existence of a large number of indigenous maize cultivars suitable for different altitudes and purposes. During 2003–2004, approximately 58 local

(a) (b)

Figure 2.
(a) Paheli makai and (b) Seti makai.

Figure 3.
Maize diversity grown in the Sikkim Himalaya.

Figure 4.
Kali makai, Ribdi, West Sikkim.

germplasms were collected by Indian Agricultural Research Institute at Tadong Gangtok Sikkim from four districts (now six districts) in Sikkim at different altitudes (**Figure 6**) [22].

The local germplasm collection program in Sikkim has identified a variety of indigenous maize cultivars, including white kernel maize known as *seti makai*, yellow kernel maize referred to as *pahenli makai*, orange to red kernel maize known as *rato makai*, and purple kernel maize named *baiguney makai*. There are also high-altitude maize types such as *Lachung makai*, *sehrung*, and *tempo ringing*, as well as *sano makai* which is a type of popcorn. The *seti makai* has variations and is grouped as

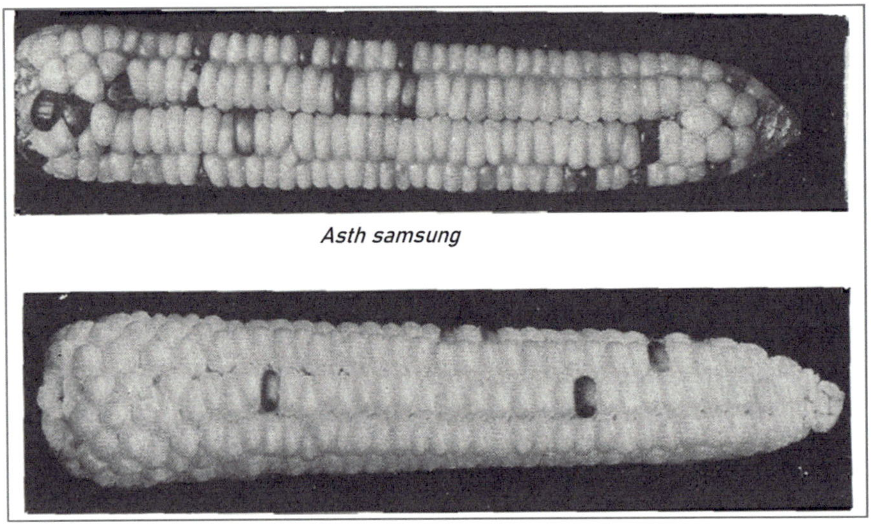

Figure 5.
Tsungrhu a sub-race of Asth Samsung (picture scanned from Singh, [17]).

Figure 6.
Diversity of different traditional varieties of maize.

Seti Makai-1 to 4, while *Pahenli makai* has six sub-types designated as *pahenli makai* 1–6. *Rato makai* has four sub-types named as *rato makkai* 1–4, and *sanu makai* has three sub-types referred to as Sikkim Popcorn 1–3. Finally, *baiguney makai* has two sub-types: *sano baiguney* and *thulo baiguney*. However, there are also many nondescript cultivars without specific names [22].

Over the past 30 years, the Sikkim Himalayan regions have documented a multitude of intriguing local variations of maize, each possessing distinct and specific traits, including *Badam-topo* (popcorn variety), *Baiguney Makai* (purple, soft variety), *Bancharey-makai* (high-altitude maize with yellow, flint kernels), *Chakhou chujak* (aromatic, soft, and sticky maize), *Chepti-makai* (white dent-type maize with distinct kernels), *Chujak* (aromatic popcorn variant), *Darikincho* (small, yellow, hard kernels), *Fingdong* (aromatic popcorn with a distinctive flavor), *Gadbade-makai* (white kernels with occasional purple flint kernels), *Kaali-makai* (dark purplish-black colored maize), *Kholakitti* (sticky variety), *Kuchung-dari* (orange-colored popcorn with flint kernels), *Kuchung takmar* (mix of yellow, white, purple, and red kernels with flint texture), *Kukhurey-makai* (dwarf, high-altitude maize), *Kukidolong-makai* (flint variety), *Lachung-makai* (variety with multiple colors and tolerance to cold conditions), *Nepali Sappa* (unique with three cobs per plant), *Pahenli-makai* (light dent type), *Pahenli-makai* (yellow/orange flint kernels), *Poorvi Botapa, Phensong-makai* (cob length of up to 30 cm), *Putali-makai* (multi-colored appearance), *Rato-makai* (dark red maize variant), *Sathiya-makai* (early-maturing type), *Seti-makai* (white, soft variety), *Tanee-makai* (popcorn type), *Tista Mehdi-makai* (flint variety), *Asthra Samsung*, *Tsungrhu*, *Sikkim Primitive I*, and *Sikkim Primitive II*. These maize variations exemplify the abundant diversity and unique characteristics found in the region.

The local germplasm displayed significant variation in cob orientation, cob size, kernel color, leaf orientation, silk color, height at which ear arises, cob length, number of kernels per row, and kernel yield per plant. '*Lachung makai*' exhibited para-mutation (multi-colored cob) and showed tolerance to cold weather. Some of the *seti makai* and *pahenli makai* had thick husk coverage and oblong cob orientation, which impart resistance against ear rot in the rainy season. The high-altitude maize *tempo ringing* matures in 85–90 days in mid-hills when other maize did not complete silking, indicating an extraordinary early maturity trait in mid and high-altitude maize. Local germplasm such as *murali makai*, *tempo ringing*, and *seti makai* are being utilized in the ongoing breeding programme at the Indian Council of Agriculture Research (ICAR) Sikkim Centre. In a nutshell, the maize genetic resources in Sikkim are rich, which is why the Sikkim Himalaya is considered a secondary centre of diversity for maize (**Figure** 7).

(a) (b) (c) (d)

Figure 7.
(a) Gadbadey makai, (b) Rato makai, (c) Thulo-baiguney makai and (d) Raato makai.

Lachung makai, characterized by its paramutation trait (multi-colored cob) and tolerance to cold weather, stands out among the local varieties. This maize variety is grown in a high temperate agroclimatic range between 2200 and 2600 m elevations. Additionally, other notable landraces such as Sikkim Primitive, *Tirap*, *Naga Sahyup* (Arunachal Pradesh), and *Tistamehdi* (Sikkim) have been previously collected and evaluated by other researchers, amounting to over 200 landraces studied (**Figure 8**) [10].

Murali makai is a crop that grows well in high altitudes (between 1000 and 1800 meters above sea level) and can withstand moisture-stress conditions, although it has slow vegetative growth. It has the potential to contribute adaptability and multiple cob-bearing traits to otherwise desirable varieties of maize grown in mid and high hills. Unfortunately, this rare genotype is gradually disappearing from cultivation and is considered an endangered cultivar (**Figure 9**).

5. Sikkim primitive maize variety

The economy of North-eastern India relies primarily on rice cultivation; however, in Sikkim, maize is the dominant crop. This underscores the crop's significance, although there has been a recent shift toward more profitable crops. Maize originated in Mexico, but India harbors considerable genetic diversity. The maize landraces in Sikkim can be classified into four groups based on their historical origin: primitive, advanced or derived, recent introductions, and hybrid races. The primitive group

(a) (b)

Figure 8.
(a) Sikkim Primitive 1, Murali makai grown in Dzongu (photo courtesy: Dawa Lhendup Lepcha) and (b) Tempo Rinzing, Lachung, North Sikkim (photo courtesy: Hishey Lachungpa).

(a) (b)

Figure 9.
Murali makai (a) Tinvong, Dzong North Sikkim, and (b) Amba, East Sikkim.

comprises several races of popcorn, including *Poorvi Botapa*, *Tirap Nag-Sahypung*, *Arun Tepi*, and *Alok Sapa*, which are distributed in the Eastern Himalayas, including Sikkim, at elevations ranging from 600 to 2000 m. The most primitive race, *Poorvi Botapa*, exists in pure form in North Sikkim, where it is known as *murali makai*. Two forms of *murali makai* are present, with different kernel colors (purple and yellow kernel maize), and designated as Sikkim Primitive-1 (purple kernel type) and Sikkim Primitive-2 (yellow kernel type). Notable varietal characteristics of *murali makai* include its prolific ear-bearing ability (with 3–5 cobs per plant, a rare trait in commercial maize), the presence of style remnants on the mature kernel, small cob size (8 cm long and 6 cm girth), and approximately 100 small kernels per cob. Despite the small kernel size, the popping efficiency is high, with a 100 kernel weight of 9.5 g (**Figures 10** and **11**).

Sikkim Primitive maize strains, namely Sikkim Primitive-1 (purple colored), and Sikkim Primitive-2 (yellow colored), were found to differ from the primitive Mexican races [5, 10]. Some of the local varieties, such as *Seti* and *Pahenli*, exhibit a thick husk coverage and an oblong cob orientation, which provide resistance against ear rot during the rainy season. Extra-earliness, a rare trait in mid and high-altitude maize, is observed in the high-altitude maize variety called *Tempo Ringing*, which reaches maturity in 85–90 days in mid-hills, outpacing other maize cultivars in silking completion. Fascinating local germplasm with promising traits, such as *Murli*, *Tempo Ringing*, and *Seti*, are currently being utilized in an ongoing breeding program at the ICAR, Sikkim Centre, Tadong (**Figure 12**) [22].

Figure 10.
Sikkim Primitive 1, grown in Dzongu, North Sikkim; Sikkim Primitive 2, grown in Dzongu, North Sikkim.

Figure 11.
Murali makai (Sikkim Primitive 2) grown in Sikkim.

5.1 Sikkim Primitive: a distinct maize landrace

In a study conducted by Sharma et al. [23], a diverse set of 48 maize landraces/locals sourced from various agroecologies across India were selected for analysis. Among these, 8 accessions were identified as "Sikkim Primitives", which were collected from

Figure 12.
Tempo rinzing maize variety of Lachung, North Sikkim (photo courtesy: Hishey Lachungpa).

different villages in Sikkim in November 2005. The remaining 40 accessions comprised 21 landraces from Northeastern Hill (NEH) regions (excluding Sikkim Primitives), 4 from non-NEH tribal hill regions, and 15 from the plains of India. This diverse set of landraces provided an opportunity to understand the genetic and phenotypic diversity present in maize germplasm in India. This study aimed to characterize maize landraces in India using both intensive phenotypic and molecular analyses. Through multilocation analyses of selected accessions, the study identified several promising landrace accessions that could be potentially useful in breeding programs. The utilization of SSR markers analyzed through DNA Sequencer technology by Sharma et al. [23] enabled an effective differentiation of accessions. Additionally, the Sikkim Primitives landrace accessions were found to be distinct at both the phenotypic and molecular levels when compared to other landrace accessions in India. These findings highlight the importance of conducting detailed characterization of germplasm collections to fully exploit their genetic potential.

5.2 Novel quantitative trait loci of '*murali makai*'

The maize landrace 'Sikkim Primitive' is known for its high productivity, producing five to nine ears per plant. However, the genes responsible for this trait had not been identified. Prakash et al. [24] conducted a study on 'Sikkim Primitive' maize landraces. They found a prolific inbred line called 'MGUSP101' which was developed and crossed with two non-prolific inbred lines, HKI1128 and UMI1200. Two F2:3 populations were evaluated across three locations. The number of ears per plant varied from 1.35 to 5.38 in the MGUSP101 × HKI1128 population. Using bulked-segregant analysis and targeted QTL mapping, a major QTL (bin: 8.05) that explained 31.7% of phenotypic variation was identified among 145 F2:3 individuals. The QTL was validated in 138 F2:3 individuals of MGUSP101 × UMI1200, and it explained 29.2% of phenotypic variance at the same interval. The novel QTL was designated as 'qProl-SP-8.05', and six candidate genes responsible for prolificacy were identified. This finding provides an opportunity to use marker-assisted selection to introgress the novel QTL for prolificacy in elite maize, and it represents the first report of the identification of the locus governing prolificacy in 'Sikkim Primitive' (**Figure 13**).

The local maize variety called *murali makai* in Sikkim, also known as Sikkim Primitive maize, possesses exceptional prolificacy and popping efficiency, making it a significant genetic resource. However, its population has suffered a severe decline, and its conservation has been overlooked, putting it at risk of extinction. To initiate the conservation and revival of this variety, Kapoor et al. [25] conducted a study that involved a characterization and documentation process. This involved assessing 31 morphological traits at different growth stages and conducting molecular characterization using simple-sequence repeat (SSR) markers.

5.3 Morphological and molecular characterization of Sikkim maize

'Sikkim Primitive' is a unique landrace discovered in the North-Eastern Himalayan region of India, particularly in the province of 'Sikkim' [24]. This landrace stands out due to its distinct ear and fruiting morphology, which closely resembles that of maize, setting it apart from teosinte where grains are enclosed in a hard fruit case or 'cupule' [21]. Detailed morphological studies conducted by Sachan and Sarkar [13] confirmed the close relationship of 'Sikkim Primitive' with maize. The designation of this landrace as 'primitive' is justified by its remarkable traits, including high prolificacy, sensitivity to photoperiod, small popcorn-like kernels, and the ability to produce abundant pollen. The unique characteristics exhibited by 'Sikkim Primitive' make it an exceptional variety deserving of further investigation and conservation efforts [24].

The plants of the Sikkim Primitive maize, locally known as *murali makai*, were observed to display remarkable prolificacy with each plant bearing 5–6 cobs. They also exhibited excellent popping capacity and several other distinctive traits. The plants

Figure 13.
Murali makai stored over the fireplace at Tinvong, Dzongu, North Sikkim.

were tall, and their stems were thin with loose drooping tassels. The base of the glumes and brace roots displayed anthocyanin colouration. The cobs were medium-sized and carried small seeds with low test weight, weighing 87.90 g [25].

Molecular characterization using 22 SSR markers demonstrated the amplification of unique amplicons, ranging from 100 to 800 bp. Of these markers, bnlg1083, umc1353, umc1128, bnlg1017, bnlg2077, umc2298, and umc2373 displayed distinct amplification patterns. The characterization of these traits and the molecular markers will be beneficial in utilizing Sikkim Primitive maize for genetic improvement and maintaining genetic purity [25].

6. Maize-based farming and its significance

In the region encompassing Sikkim and the Darjeeling Himalayas, maize cultivation is of paramount agricultural significance. This area has emerged as an optimal location for maize farming due to its favorable climatic conditions and suitable soil characteristics. Farmers in this region dedicate substantial land to maize cultivation, making a significant contribution to the overall agricultural output. The agroclimatic conditions within an elevation range of 200 to 2700 meters are particularly suitable for cultivating a diverse range of maize varieties. This includes as many as 26 traditional maize landraces and approximately 15 different hybrid and certified maize varieties, showcasing the rich agricultural diversity of the region.

Farmers in Sikkim and the Darjeeling Himalayas employ a blend of traditional and modern farming techniques to cultivate maize. They adhere to best agricultural practices, encompassing appropriate land preparation, meticulous seed selection, precise planting methods, and effective pest management strategies. These practices are implemented to maximize yield and maintain the overall health of the crop.

The maize production, productivity and area under cultivation in Sikkim from 2017 to 2018 to 2021-2022 is given in **Figure 14**. The area dedicated to maize cultivation in this region during 2021–2022 in Sikkim was 38,458 ± 580 hectares. This substantial figure underscores the significant role of maize farming in the regional agricultural landscape. Additionally, the region boasts an average annual maize production of around 67,692 ± 1243 tonnes [27]. This considerable output highlights the success and potential of maize cultivation in meeting the food requirements of the local population and beyond (**Figure 15**).

Between the years 1981 and 1990, there was a notable increase in maize cultivation in Sikkim, with a 27.6% expansion in acreage. During this period, there was a substantial increase of 174% in total production and 111% in yield per unit. These advancements were primarily attributed to the introduction of high-yielding varieties (HYVs) in Sikkim. However, the subsequent decade from 1991 to 2001 saw a decline in the growth rate, with only a 31.7% increase in total production and a 22.3% increase in yield per unit recorded. Interestingly, despite having more choices of HYVs available during the latter period, the sustained growth seen in the previous decade could not be maintained [28]. However, according to reports from the Agriculture Department of the Government of Sikkim, there has been a significant decrease in the cultivated area from 2012 to 2022, with a reduction of 1892 hectares. While productivity and production levels have remained relatively stable during this period.

Maize cultivation in the region encompasses contributions beyond economic and food security aspects and assumes a pivotal role in supporting the livelihoods of farming communities. Its practice generates employment opportunities, facilitates

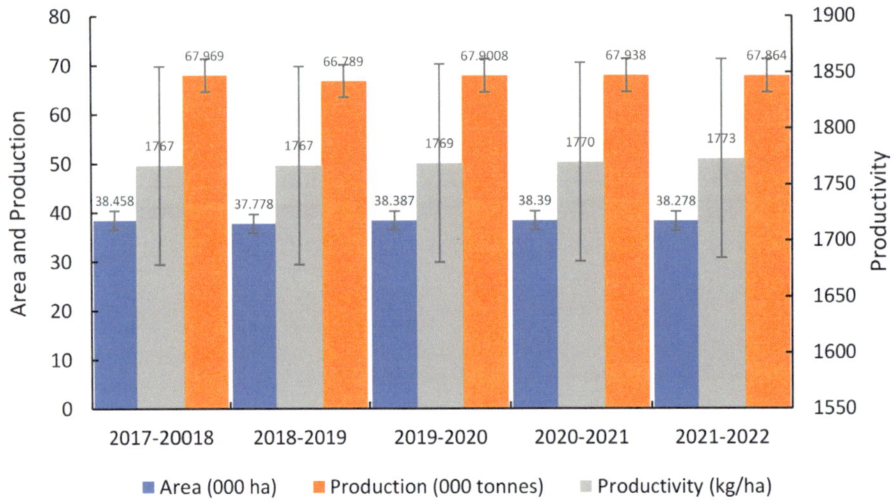

Figure 14.
Area, production and productivity of maize in Sikkim, India from 2017 to 2022 (data compiled from [16, 26, 27]).

Figure 15.
A maize production farm at Martam, near Gangtok.

rural development, and strengthens the overall resilience of the agricultural sector. Since 2012, a total of 17 high-yielding varieties (HYVs) and hybrid varieties have been introduced in Sikkim. Alongside these introductions, hybrid and certified maize seeds obtained from external sources, such as Pusa Vivek (QPM 9 (Improved), NAC-6004 composite, 33 M66, 66 K99, JKMH-1701, have also been adopted in Sikkim [27]. Based on the experiences shared by farmers in various regions of Sikkim, it has been observed that nine prominent pests have infested the high-yielding varieties (HYVs) and hybrid maize varieties. These pests are the stem borer (*Chilo partellus*), cutworms (*Agrotis ipsilon*), armyworms (*Mythimna separata*), semi-loopers (*Plusia signata*), and cob borers (*Stenchroia elongella*). It is noteworthy that the indigenous landraces of maize exhibit tolerance to many of these pests [29].

A very high intensity of post-harvest infestations have been recorded by insects such as the maize weevil (*Sitophilus zeamais* Mostsch.) and grain moth (*Sitotroga cerealella* Oliv.) in the HYVs and hybrid maize varieties, which have been identified as major storage insect pests. While pest infestation in the indigenous varieties was significantly 80% less as compared to improved varieties.

Continuous efforts are underway to enhance maize production in Sikkim and Darjeeling Himalayas. These initiatives encompass the promotion of improved farming practices, the introduction of high-yielding maize varieties, and the implementation of sustainable agricultural techniques. These endeavors aim to ensure long-term food security, augment farmers' income, and foster comprehensive agricultural development in the region. The maize cultivation in Sikkim has proven to be a vital agricultural activity, contributing to food security, bolstering economic growth, and enhancing the well-being of local communities (**Figure 16**).

6.1 Maize inter-cropping

The study compared the productivity of intensified cropping systems to traditional maize-fallow systems in the rainfed region of Sikkim Himalayas. The results showed that intensified cropping systems, such as maize (green cobs)–urd bean–buckwheat and maize–rajmah, had significantly higher maize grain equivalent yield and system production efficiency compared to the maize-fallow system. The maize (green cobs)–urd bean–buckwheat system demonstrated the highest relative production efficiency and land use efficiency due to its longer crop duration, while the maize-fallow system had the lowest land use efficiency. These findings highlight the potential of intensified cropping systems to increase agricultural productivity and land use efficiency in the region. The study suggests that promoting the adoption of appropriate intensified cropping systems could contribute to enhancing food security and farm productivity in the rainfed areas of Sikkim Himalayas.

Figure 16.
Maize varieties stored under the roof.

6.2 Economics of maize inter-cropping

In an economic analysis conducted by Sing et al. [30], it was found that the maize (green cobs)–urd bean-buckwheat cropping system had the highest net return of 303,000 INR per hectare and a benefit-cost ratio of 2:6, which was significantly better than the maize–rajmah system. Conversely, the lowest return and benefit-cost ratio were observed in the maize–fallow cropping system. In terms of relative economic efficiency (REE), which is a comparative measure of economic gains over the existing system, all the intensified systems had higher economic gains than the maize-fallow system. Moreover, the maize (green cobs)-urd bean-buckwheat cropping system had the highest system profitability of 831 INR per hectare per day, while the maize–fallow system had the lowest system profitability of 238 INR per hectare per day (**Figure 17**).

Employment generation is a key indicator when assessing the sustainability of cropping systems. The data showed that the intensified systems added to the employment generation and generated more employment than the maize-fallow cropping system, which only generated 106 man-days per hectare to harvest the final produce. The relative system employment generation efficiency (REGE), which measures the additional man-days required for a diversified system compared to the existing system, revealed that all the intensified systems had higher employment generation ability than the prevailing system in the region [30]. Among the cropping sequences, the cultivation of maize (green cobs)-urd bean-buckwheat resulted in the maximum REGE of 168%, followed by the maize-rajmah system.

6.3 Post-harvest storage management

Once the corn cobs are harvested and sun-dried, a traditional method of preservation is employed in the mountains of Sikkim Himalaya. To accomplish this, four to six cobs are carefully selected and their husks are tied together. In preparation, the outer rough and aged husks are removed, while some of the inner husks are torn

Figure 17.
Maize intercropping with a variety of other crops.

without completely separating them from the shank, forming what is known as a husk tail. These husk tails are then used to tie together a minimum of four cobs, creating a bundle known as "*Jhutta*." This bundling technique helps to keep the cobs secure and protected from external elements, ensuring their quality and freshness during storage.

Maize farmers in Sikkim employ organic cultivation methods and employ effective post-harvest storage techniques, abstaining from the use of chemical pesticides and adopting improved seed bins. This sets them apart from farmers in other states of India who rely on chemical interventions. Pest management in these regions follows a meticulous decision-making process aimed at controlling pests in a cost-effective manner.

Sun drying has traditionally been the primary method employed for pest management practices in this region. Additionally, several bio-rational plant products have shown superior efficacy in controlling maize weevils. These include Sweet flag (*Acorus calamus*), Neem oil (*Azadirachta indica*), Neem seed powder, Timur (*Zanthoxylum armatum*), and Titepati (*Artimesia vulgaris*). These bio-rational plant products have proven to be effective alternatives to synthetic pesticides. Furthermore, approximately 10% of farmers utilize barriers such as leaves of *Pinus roxburghii*, *dalle kuro* (*Urena lobata*), *babio* (*Eulaliopsis binata*), as well as red and white soil, to prevent rat invasions. These preventive measures contribute to pest control and minimize crop damage. By employing organic methods, utilizing bio-rational plant products, and implementing physical barriers, maize farmers in Sikkim demonstrate their commitment to sustainable pest management practices. These approaches not only ensure the production of high-quality maize but also contribute to the preservation of the environment and the health of consumers (**Figure 18**).

Following the tying of the corn cobs, the next step in the storage process is finding suitable locations in the house. One common method is to store them on "*Thangro*," which is a structure made of a vertical pole topped with a rooftop. Another option is to hang the tied cobs under the eaves in a structure known as *Bardali*. These storage practices ensure that the cobs are kept off the ground,

Figure 18.
Storing Pahenli makai in a Thangra.

preventing moisture absorption and minimizing the risk of pests and rodents. By employing the *Kunyo* or *makaiko-haar*, the communities in this region can maintain the quality and availability of their corn supply throughout the year. These storage arrangements provide safe and elevated spaces for the corn cobs, protecting them from moisture, pests, and rodents that can compromise their quality. By employing these storage methods, the communities ensure that the stored maize remains preserved and readily available for consumption throughout the year (**Figures 19** and **20**).

Figure 19.
Storing maize in a Thngra.

Figure 20.
Storing maize in a rope alongside of a house.

7. Soil fertility and crop management

Soil fertility has been identified as a significant constraint affecting maize production throughout all districts of Sikkim. The availability and quality of manure/compost have been recognized as crucial inputs for sustaining and improving soil fertility. Farmers have expressed concerns regarding limited access to an adequate supply of manure/compost, primarily due to the scarcity of high-quality fodder for their livestock. The quality of compost has shown considerable variation, indicating that many farmers have not yet adopted improved compost management practices.

In relatively more accessible areas, farmers have resorted to supplementing manure/compost with organic fertilizers provided by the Department of Agriculture. However, the use of fertilizers has been predominantly limited, raising concerns about potential deficiencies in other essential nutrients, particularly phosphorus. The organic inputs commonly used by farmers have been found to be insufficient sources of phosphorus. Soil erosion has emerged as a significant challenge, resulting in substantial loss of productive topsoil, especially in fields with sloped terrain and experiencing intense monsoon rainfall.

Indigenous traditional knowledge and practices associated with soil fertility maintenance include *in situ* farm manuring, mulching, bio-composting, green manuring, livestock ranching, cultivation of nitrogen-fixing plants, land fallowing, and litter decomposition. These practices can contribute to the improvement and sustainability of soil fertility in maize cultivation.

8. Uses and culinary practices

Maize has a long-standing significance as a dietary staple in the hilly regions of Sikkim, where it is deeply ingrained in various traditional culinary practices. In these practices, the dried maize seeds are commonly utilized as the primary ingredient for the preparation of popcorn. The green cobs, carefully harvested at the milk stage, undergo a grinding process to produce a versatile food product known as *phyaplo*. Moreover, these green cobs are also employed in the creation of local bread varieties (**Figure 21**).

Another notable utilization of maize involves the light beating of maize to separate the husk and the kernel. The kernels are then soaked in hot water for 24 hours and is roasted lightly on fire and the hot kernels are beaten in wooden *oklhi—musli* to cup shape structure referred to as *chyadung/chadung*. This form adds diversity to the range of maize-based food items. Additionally, maize flour, obtained by grinding the maize kernels, is transformed into a dough-like consistency, which is then spread evenly onto banana leaves. The dough-laden banana leaves are then subjected to the radiant heat generated by burning wooden coal in a traditional fireplace called *Agenu*. This process, known as *Bhungrey-roti*, facilitates the roasting and cooking of the maize flour, yielding a unique and culturally significant culinary creation. In a traditional culinary practice, freshly harvested maize seeds are subjected to pounding using either a *Dhiki* (a wooden pounding tool) or an *Okhli-musli* (wooden mortar and pestle). This pounding process results in the production of beaten rice, which is a popular food item. The beaten rice is commonly consumed with hot milk, creating a nourishing and satisfying meal. This method of processing maize seeds into beaten rice demonstrates the resourcefulness and utilization of maize as a versatile food source in the region.

The utilization of maize in these scientific cooking techniques showcases the rich culinary heritage and cultural importance of maize in the hilly regions of Sikkim.

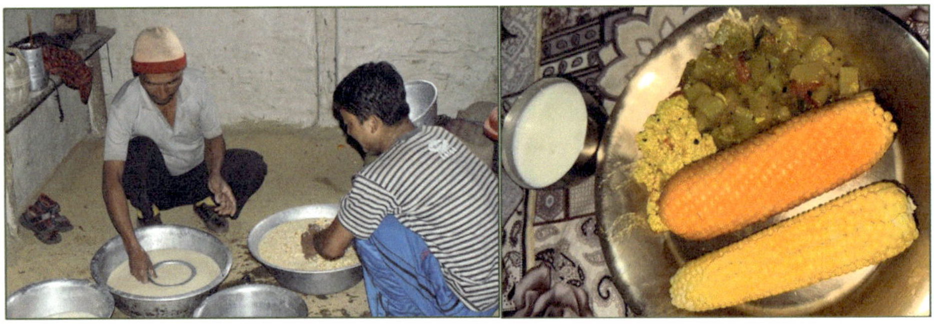

(a) (b)

Figure 21.
(a) Fermented Jarrnd preparation out of maize kernels by Mangar community of Karjee, West Sikkim and (b) boiled maize with Indian pumpkin curry, cheese pickle and buttermilk as dinner.

In addition, young green cobs are often consumed after being roasted or boiled. Maize grains are partially ground, and a combination of rice with an equal ratio of *makkaiko chamal* (maize grains) and powdered maize grain starch is used to prepare *makkaiko pitho*, and *chyakhla* (coarse maize rice) or *saraulo* (fine maize rice). Roasted seeds are grinded to powder known as *'champa'* or *'Saatu'*. Sometimes, *champa* preparation includes a blend of wheat or barley. Maize seeds are boiled and a fermented product called *Makaiko-Jarnd* is prepared. Wines are also prepared using fermented maize products.

9. Conclusion

Primitive maize cultivars in the Northeastern Himalayas of India have generated interest among researchers for their high prolificacy (4 to 8 ears/plant) and their link to the origin and evolution of maize [31]. Sachan and Sarkar [14] concluded that Sikkim Primitive maize is equivalent to pre-Chapalote, pre-Nal-Tel, and prehistoric wild corn of Mangelsdorf. Besides being of interest for origin and evolution, Sikkim Primitives serve as a valuable source of prolificacy, pest resistance, and drought tolerance, given their resilience against natural challenges.

To enhance farm productivity and food security in the rainfed Eastern Himalayan region, it is necessary to intensify the existing maize-fallow system by incorporating more crops per unit area. This requires careful crop selection and increased productivity, especially in rainfed ecosystems. However, intensified production systems also demand higher energy and other inputs. Therefore, focusing on sequential cropping in mono-cropped areas and crop intensification is crucial for improved agricultural production. Vertical growth, by increasing productivity per unit of land, offers an alternative to expanding horizontally due to limited space.

Genetic characterization and improvement of local and indigenous maize cultivars are pivotal in addressing their limitations. In Sikkim, the rapid introduction of high-yielding and hybrid varieties has led to the decline of traditional landraces in agroecosystems. Thus, conserving these landraces in the National Bureau of Plant Genetic Resources (Indian Council of Agricultural Research) is imperative for future research, registration, and protection under the Protection of Plant Varieties and Farmers' Rights Act 2001.

Diversification and intensification of maize-based systems, through traditional crop management practices, can enhance profitability and resource use efficiency in both irrigated and rainfed ecosystems of the Sikkim Himalayas. Diversifying monocropped maize areas improves the profitability and productivity of organic agriculture, sustaining the livelihood security of organic growers in Sikkim. This approach supports the development of a more sustainable agricultural system, contributing to food security and livelihood enhancement.

The Eastern Himalayan region of India is widely recognized as a significant secondary centre of maize diversity, housing a diverse array of genetic resources. Maize cultivation in this region spans altitudes from 300 to 2500 masl. A field investigation conducted between 2010 and 2022 employed a combination of traditional knowledge and scientific methods to characterize maize morphologically. A comprehensive literature review documented the diversity of maize cultivation in the Sikkim Himalaya since the 1960s. The primary objective was to assess the number of indigenous maize varieties cultivated in the Sikkim Himalayan region approximately 60 years ago until 2023.

Categorizing maize landraces provides valuable insights into the diversity and classification of maize varieties, contributing to our understanding of maize genetic resources. The study also involved the collection of *in situ* germplasm in farmers' fields to conserve cultivated plant diversity. Prioritization at the species level and in specific geographic regions was necessary due to the dynamic demand for germplasm. The primary objective was to assess the risk of extinction or erosion of specific maize germplasm, crucial for cultural and ecological preservation. The study aimed to safeguard valuable genetic resources, ensuring their availability for food security, livelihoods, climate resilience, and research. Furthermore, the study focused on improving landraces and increasing access to high-quality germplasm.

During the study, as many as 31 different landraces of maize were identified, highlighting the need for comprehensive research on the genetic characterization of each variety, which remains unexplored in the region. Notably, *Murali Makai, Seti Makai, Pahenli Makai, Rato Makai, Baiguney Makai, Gadbadey Makai, Tempo-Rinzing*, and *Lachung Makai* exhibited excellent adaptation to local environments ranging from 300 to 2500 m. These landraces displayed variations in agronomic and quality traits, as well as resistance to biotic and abiotic stresses. Their adaptability has resulted from natural and artificial selection. Conserving these landraces is crucial for future food security and sustainable agricultural practices.

Genetic research on Himalayan maize is critical for unraveling the genetic basis of unique traits, including agroclimatic adaptability, disease and pest resistance, and abiotic stress tolerance. This knowledge is essential for breeding superior maize varieties with enhanced productivity, nutritional value, and resilience. Additionally, studying the genetic diversity of Himalayan maize aids in conserving valuable landraces and heirloom varieties, which possess traits absent in commercial cultivars, thus ensuring long-term food security and farming community resilience. Genetic research facilitates the identification and utilization of genes and markers associated with desirable traits, enabling the development of molecular breeding techniques like marker-assisted selection and genetic engineering for targeted maize improvement. Furthermore, understanding the genetic diversity and population structure sheds light on maize's evolutionary history, migration patterns, and contributes to broader scientific knowledge of crop domestication and genetic dynamics. This research supports the formulation of sustainable agricultural policies and strategies by providing evidence-based data on crop improvement, seed systems, biodiversity conservation, and farmer rights. Ultimately, the genetic research on Himalayan maize is pivotal for advancing our understanding of unique traits, conserving genetic resources, enhancing breeding programs, and fostering sustainable agricultural practices, thereby promoting food security, resilience, and the well-being of farming communities.

Acknowledgements

This study was supported by SECURE Himalaya project, a collaborative initiative between the Ministry of Environment, Forest and Climate Change, Government of India, and the United Nations Development Programme in India, with the financial support from the Global Environment Facility. We are highly grateful to The Mountain Institute India and Sikkim Biodiversity Board for facilities. We acknowledge the knowledge of the local and indigenous people i.e. Nepalese, Bhutia and Lepcha of six districts of Sikkim for their valuable knowledge on maize cultivation and conservation of maize genetic diversity.

Conflict of interest

The authors declare no conflict of interest.

Author details

Ghanashyam Sharma[1,2]* and Bharat Kumar Pradhan[1,2]

1 The Mountain Institute India, Gangtok, Sikkim, India

2 Sikkim Biodiversity Board, Forest and Environment Department, Gangtok, Sikkim, India

*Address all correspondence to: sharmaghanashyam@yahoo.com

References

[1] Prasanna BM, Cairns JE, Zaidi PH, Beyene Y, Makumbi D, Gowda M, et al. Beat the stress: Breeding for climate resilience in maize for the tropical rainfed environments. Theoretical and Applied Genetics. 2021;**134**:1729-1752

[2] Iltis HH, Doebely JF. Taxonomy of *Zea* (Gramineae) II. Subspecific categories in the *Zea mays* complex and generic synopsis. American Journal of Botany. 1980;**67**(6):994-1004

[3] Beadle GW. The ancestry of corn. Scientific American. 1980;**245**:112-119

[4] Doebley JF, Iltis HH. Taxonomy of *Zea*, I. A subgeneric classification with key to taxa. American Journal of Botany. 1980;**67**(6):982-993

[5] Mukherjee BK, Gupta NK, Singh SB, Singh NN. Meteroglyph analysis of Indian and exotic maize. Euphytica. 1971;**20**:113-118

[6] Manglesdorf PC. Corn: Its Origin, Evolution and Improvement. Cambridge, Massachusetts, USA: Harvard University Press; 1974

[7] Wilkes HG. Maize in India. Maize Genetics Cooperation Newsletter. 1981;**55**:13-15

[8] Dowswell CR, Paliwal RL, Cantrell RP. Maize in the Third World. Boulder, Colorado, USA: Estview Press; 1996

[9] Marszewski T. The problem of the introduction of primitive maize into Southeast Asia. Folia Orient. 1978;**19**:127-163

[10] Prasanna BM, Sharma L. The landraces of maize (*Zea mays* L.): Diversity and utility. Indian Journal of Plant Genetic Resources. 2005;**18**(2):155-168

[11] Anderson E. Maize in the New World. In: Wilson CM, editor. New Crops for the New World. New York, USA: Macmillan; 1945

[12] Dhawan NL. Sikkim primitive maize. MNL. 1964;**38**:69-70

[13] Sachan JKS, Sarkar KR. Plant type of 'Sikkim Primitives'. Maize Genetics Cooperation Newsletter. 1982;**56**:122-124

[14] Sachan JKS, Sarkar KR. Discovery of Sikkim primitive precursor in the Americas. Indian Journal of Genetics. 1986;**46**:153-161

[15] Tanksley SD, McCouch SR. Seed banks and molecular maps: Unlocking genetic potential from the wild. Science. 1997;**277**:1063-1066

[16] APR. Annual Progress Report 2021-2022. India: Government of Sikkim, Department of Agriculture; 2022

[17] Singh B. Races of Maise in India. New Delhi, India: Indian Agricultural Research Institute. Printed in India by V. Ananda Krishna Chettty, at Hoe and Co.; Madras and Published by SS Grewal, Under-Secretary to the Indian Council of Argiculture Research, New Delhi; 1977

[18] Mangelsdorf PC. Reconstructing the ancestor of corn. Science, USA. 1958;**128**:1313-1320

[19] Mangelsdorf PC, MacNeish RS, Galinat WC. Domestication of corn. 1964;**143**:538-545

[20] Bhat BK, Dhawan NL. Effect of cytoplasm on quantitative character of maize. Indian Journal of Genetic Plant Breeding. 1969;**29**:321-326

[21] Sharma L, Prasanna BM, Ramesh B. Analysis of phenotypic and microsatellite-based diversity of maize landraces in India, especially from the north east Himalayan region. Genetica. 2010;**138**:619-631

[22] Rahman H, Karuppaiyan R. Agrobiodiversity of Sikkim. In: Arrawatia ML, Tambe S, editors. Climate Change in Sikkim: Patern, Impact and Initiatives. Government of Sikkim, Gangtok: Department of Information and Public Relations; 2011. pp. 403-428

[23] Sharma L, Singh NK, Guleria SK, Ramesh JS. Molecular distinctness from other maize landraces in India. In: Asian Regional Maize Workshop, 10. Maize for Asia-Emerging Trends and; Technologies. Proceedings of The Asian Regional Maize Workshop; Makassar, Indonesia; 20-23 October, 2008 (p. 250). CIMMYT; 2008

[24] Prakash NR, Zunjare RU, Muthusamy V, Chand G, Kamboj MC, Bhat et al. Genetic analysis of prolificacy in "Sikkim Primitive"-A prolific maize (*Zea mays*) landrace of North-Eastern Himalaya. Plant Breeding. 2019;**138**(6):781-789. DOI: 10.1111/pbr.12736

[25] Kapoor C, Singh S, Awasthe RK, Shankar SM, Pattanayak A, Raj C, et al. Harnessing potential of maize (*Zea mays*) genetic review for exploring yield and yield-related traits under organic farming in hill region. Plant Genetic Resources: Characterization and Utilization. 2022;**20**:425-433

[26] APR. Annual Progress Report 2020-2021. India: Government of Sikkim, Department of Agriculture; 2021

[27] HBAS. Hand Book of Agriculture 2019-2020. Agriculture Department, Government of Sikkim, India: Agriculture Census and EARAS Unit; 2020. p. 84

[28] AR. Annual Report 2011-2012. Krishi Bhawan, Gangtok Sikkim, India: Department of Agriculture, Government of Sikkim; 2012

[29] Singh R, Kalita H, Kapoor C, Gopi R, Babu S, Das SK. Maize (*Zea mays* L.). In: Awasthe RK, Pradhan Y, Bhutia K, editors. Handbook of Organic Production in Sikkim. Jointly Published by Sikkim Organic Mission, Government of Sikkim, Gangtok Sikkim, and ICAR Comple for NEH Region. Sikkim Centre, Gangtok Sikkim India; 2014. p. 408

[30] Singh R, Babu S, Avasthe RK, Yadav GS. Crop Diversification and Intensification for Enhancing Livelihood Security in Sikkim. Tadong, Gangtok, Sikkim, India: Published by ICAR-National Organic Farming Research Institute; Ministry of Agriculture and Farmers Welfare, Government of India; 2018

[31] Mani VP, Jotshi HC, Koranne KD. Preliminary observations on drought tolerance in Sikkim primitive maize. Maize Genetics Cooperation Newsletter. 1987;**61**(2)

Chapter 2

Teachings of Tatéi Niwetsika: Native Maize from Northern Mexico

Cyndy Garcia-Weyandt

Abstract

In the spring of 2021, Kalamazoo College students began the project "Tatéi Niwetsika: Planting Traditional Knowledge and Flavors," with the intention of learning about traditional agriculture, Native Maize from Mexico, food sovereignty, and the connection between Wixárika language and culture in Nayarit. In November of 2022, we harvested the first crop of Native Maize, and thus the teachings of Our Mother Corn. From a community perspective and active participation in the field of research, this chapter discusses the importance and relevance of cultivating Native seeds using traditional agriculture. The author shares from multiple perspectives the activities that link academic research and community work in Tepic and Kalamazoo. This chapter focuses on issues of traditional agriculture, the challenges of planting Native seeds of the Gran Nayar, and food sovereignties. Finally, the author grapples with the challenges of planting Native Maize in Kalamazoo and the desire to adapt the seeds in another environment. This chapter aims to highlight traditional techniques of cultivating Maize and the ceremonial aspects. Additionally, this chapter aims to define essential research methods and techniques such as community participation and social justice for more reciprocal research on issues regarding Indigenous sovereignties.

Keywords: native maize from Northern Mexico, traditional ecological knowledge, Wixárika community, Our Mother Corn, food sovereignty

1. Introduction

Native communities of the highlands of Mexico domesticated Wild Corn or *Teosinte* (*Zea mays* spp.) between 5500 and 4000 years ago. About 6000 years ago, societies across Mesoamerica began consuming Maize as well as other domesticated plants such as beans and squash [1]. The domestication of Maize (*Zea mays* L.) by Mesoamerican communities marked the beginning of sedentary life for many groups of people. Archeological records indicate that the oldest maize cobs were found in Guilá Naquitz Cave dating 6250 years ago [2].

In Wixárika ways of knowing, the oral tradition indicates that Watakame, the first farmer, received *Yuawima* ("Blue Corn Maid") from Tatéi Takutsi Nakawé

(Our Grand Mother Growth). Since their first encounter, the two (Watakame and Yuawima) have been part of the same field of relations. Watakame cultivated Yuawima and Yuawima grew more seeds in the field in exchange for offerings. In this reciprocal relationship, Watakame received a set of instructions, protocols, and principles to follow Traditional Ecological Knowledge (TEK) or local understanding of the environment and the applicability of this knowledge in daily life [3–5]. Menzies argues that "TEK is an embodied practice directly rooted in everyday livelihood activities" (88). The oral tradition such as the case of Watakame is more than a cultural framework for Wixárika knowledge. This oral tradition contains knowledge (a set of protocols and principles to interact with the environment) that people developed. In consequence, specific ways of relating with the environment developed with daily activities.

According to the oral tradition, Watakame was unable to follow the first instructions and therefore, *Yuawima* returned home. Later, Watakame made the commitment to cultivate and harvest Our Mother Corn following the initial instructions.[1] Today, many families continue following those first instructions, the protocols of coexistence, and the principles to cultivate and harvest Tatéi Niwetsika ("Our Mother Corn"). Families gather to cultivate using only the *Kuwa/Wiika* (*"coa," "pichuaca,"* or "Traditional planting stick"), an agricultural tool with a wide and thick blade, sometimes curved, inserted into a wooden handle. In the center of the field, the families create small Wixárika microcosmos to deposit offerings. The *Teiyari* ("Center of the field" or "Heart of Our Mother Corn") contains a sample of the seeds for cultivation. In the field, Wixárika families coexist with La Milpa and re-create in every cycle a series of embodied practices such as cooking, cultivation, and ceremonies to maintain and sustain a reciprocal relationship with Our Mother Corn and the five Corn Maids [7]. All the embodied practices (e.g., cultivation, harvest, ritual, and cooking) surrounding the growth of Maize are pieces of ecological knowledge that transmit the balance interaction between people and the environment.

In La Sierra Madre Occidental, Wixárika families cultivate five different variants of Native Corn (Blue, White, Pink, Yellow, and Multi-color). For example, some families from El Roble cultivate *Tsinawime* ("Multi-color"), *Tekuleti* ("Blue"), *Pipitiyu* ("White"), *Ta + rawime* ("Pink"), *Taxawime* ("Yellow"), *Yek + ri tuxa* ("Orange"), and *Tse' + ri* ("Yellow with elongated") [7]. In Y + rata, families cultivate Yuwima ("Blue Corn") in the South, *Tuxame* ("White Corn") in the North, *Ta + lawime* ("Pink Corn") in the West, *Taxawime* ("Yellow Corn") in the East, and *Tsayule* ("multi-color Corn") in the center. In the work of Victor Antonio Vidal Martínez et al., an expert on Corn in the state of Nayarit, the author reports that 13 Native species of Corn come from the state of Nayarit. Seven of them are primary species: Tabloncillo, Elotero de Sinaloa, Blando de Sonora, Bofo, Elotes Occidentales, Tuxpeño, and Vadendeño. Six of them are secondary species: Tabloncillo x Tuxpeño, Tuxpeño x Tabloncillo, Elotes Occidentales, Tabloncillo x Olotillo, Tabloncillo x Blando de Sonora, and Elotes Occidentales x Elotero de Sinaloa [8].

With the information received in TEK, the Wixárika community maintains a reciprocal relationship with Our Mother Corn and conceived Native Corn from El Gran Nayar as a relative [7]. Additionally, the community cultivates and harvests Our Mother Corn in ways that do not alter the land's natural processes. This includes slash-and-burn and polyculture. Many studies have shown that Indigenous communities continue ancestral agricultural practices that include following a ceremonial cycle

[1] Personal interview with Lemus and Zing [6].

for the cultivation of different crops [9–15]. For the cultivation of Native Corn in Nayarit, families cultivate after the first rain of the season in July. However, before the cultivation, many families gather for ceremonial purposes in four important community gatherings such as for 1) the selection of seeds, 2) the blessing of the seeds, 3) the cleansing of weeds in the field, and 4) the petition for rain. Once the families are ready to cultivate, they will go to the *Yeturita* ("Field for cultivation"). In the summer, usually at the end of June or in the beginning of July, the families prepare the seeds and the *Waxata* ("Cornfield") for cultivation and for the ceremonies involving the cycle of Corn. Like the Corn Maids and Watakame, women and men work in the field. Women handle the seeds and men work the soil and prepare the Waxata for cultivation. In ceremonies such as in the *Tatéi Niwestsika 'Etsixa* ("Cultivation Ceremony"), children perform a series of dances to embody all Corn maids. They dance to the sound of the violin and present the offerings to *Tatéi Takutsi Nakawé* ("Our Mother Growth") and Tatéi Niwetsika. Then, they exchange Corn-based offerings among themselves. After the ceremony, women continue the labor of husking, selecting the best Our Mother Corn seeds, and planning the next trip to the *Yeturita* ("Field for cultivation") to begin the season [7].

Traditional agriculture employed in the Milpa allows for the cultivation of multiple crops. According to Kremen and Miles, the diversification of farming systems regenerates the ecosystem by following techniques such as "composting, cover crops, crop rotation, absence of synthetic pesticides and fertilizers" [16]. Crops are grown together to increase biodiversity, enhance soil health, eliminate fossil fuel fertilizers and pesticides, and control erosion [17]. In many Wixárika families, they cultivate beans, squash, chilies, herbs, amaranth, medicinal plants, and flowers. Each plant companion, like the three sister systems among Anishinaabe [18] and Chakra systems in the Andes [19], serves a different role in the support of the ecosystem. For instance, Our Mother Corn provides the structure by growing vertically, so that beans can crawl to maximize the consumption of light. In exchange, beans maximize the production of oxygen and nitrogen in the soil. Squash grows closer to the soil to prevent predators and allow the other plants to grow healthier. Finally, wild medicinal plants, insects, and pollinators grow in the Milpa, so that people can benefit from their properties while consumed. In a field trial, the growth of "fava bean/maize intercrop" showed that beans facilitated the growth of maize yields by moving phosphorus that was consumed by maize [16].

2. Academic research and community work in Tepic-Kalamazoo

The main purpose of the *La Milpa Project* in Tepic-Kalamazoo is to contribute from all angles to the strengthening of the Native culture and languages of Gran Nayar. This project aims to not only facilitate workshops for students or the community but to promote actions that make the efforts of the community and their resistance to assimilation visible in Tepic. For this, the Center for International Studies (CIP) and Critical Ethnic Studies department at Kalamazoo College joined efforts with collectives, teachers, speakers of the language, and members of the research community to launch a pilot summer abroad in Tepic, Nayarit. They were invited to put together a program that could facilitate actions of community engagement. In Kalamazoo, after teaching the course "Plant Communication and Kinship," the students grew El Gran Nayar seeds for the first time. Two students from Kalamazoo College took care of the La Milpa project, while the other students traveled to Mexico to learn more about the community and the work to revitalize the Native languages of Greater Nayar. The seeds from La Milpa in our Hoop House Garden were a gift that reached Kalamazoo College with my migration and movement from Mexico to the USA between the years of 2019 and 2020. Since those seeds arrived at our school, we have documented how a community of students and teachers have learned the essentials about the care of Native Corn seeds. Together with teacher Felipa Rivera Lemus and her family from Y + rata, a Wixárika community in Tepic, the group of students from Kalamazoo has been advised and guided in the cultivation, care, and fair distribution of the seeds.

In the Fall of 2022, the Y + rata Elders arrived in Kalamazoo for a harvest festival. During their stay, they shared the way in which we all can relate to Our Mother Corn, for example, by providing food that they use as traditional offerings during a good harvest. These offerings gave *Nuestra Madre Maíz* ("Our Mother Corn") the necessary elements to close the cycle of Corn. Students learned the importance of having relationships with more-than-human persons. During the visit, we hosted a panel on the sustainable agriculture and food sovereignty to discuss the responsibility of all people in caring for the environment. This visit was significant because while the Elders taught us about Native seeds, we questioned our responsibility in caring for Native seeds in Mexico and the United States.

3. Methodology

La Milpa project follows the research methodology of Community-Based Participatory Action Research (CBPAR) [20]. According to the authors like Atalay and McCleary, "CBPAR is best understood as a decolonizing methodology intended to improve the ethics and practices of research by striving for the mutual benefit of those most affected by a particular research project through equitable, collaborative partnerships at all stages of research between researchers and community members" ([20]: 5). In the *La Milpa Project*, we know that the inclusion of community methodologies and epistemologies is important to be able to implement projects related to plants as relatives. The methodologies of the La Milpa Project since the beginning of the work have been about social change and social justice. Along with the cultivation of Our Mother Corn in agricultural practices, we acknowledge the importance of collaborating with the community on topics related to culture and language revitalization. Additionally, we follow decolonial epistemologies in the project. Decolonial practices help us understand the central role of land as our teachers. We listen to the teachings of more-than-human beings [21].

The work of linguistic revitalization is relevant to Indigenous peoples who live in the diaspora. When in connection with Traditional Ecological knowledge, we learn about the relationship between people and land. Additionally, language is a key component in understanding worldview. Understanding different ways of knowing the world helps us to demystify the relationship between body and land. For this, *La Milpa* Project works in collaboration with collectives such as the *Proyecto Taniuki* (Our Language Project) and *Yuri'Ikú* (Cultural and Gastronomic Center) to teach about the inclusion of the Wixárika community and the approach to a linguistic policy that prioritizes Indigenous pedagogies including agricultural practice.

Language in connection with agricultural practices has been a vehicle for the revitalization and strengthening of Indigenous identity. Although the workshops with *Proyecto Taniuki* are small initiatives, they have an impact on the community.

An example is the ethnolinguistic landscape project in *Lomas Bonitas* that culminated in 2022. After extensive fieldwork with the community and consultation with the community assembly, community Elders, and the general population, Kalamazoo students conducted ethnographic fieldwork to collect names for spaces and places relevant to the community. Within these spaces were public and community spaces, and topographies (mountains and hills). These spaces have meaning and social and cultural value. The members of the project included people from the community, teachers, students, and the children of the community. Together they organized namespaces in Wixárika and in Spanish. In this intergenerational work, the names, meanings, and spaces to be named were established by consensus. Thus, students from the community and outside communities conceptualize the importance of land and territory from the community's perspective.

3.1 Study area: Tepic and Kalamazoo

In the work of growing La Milpa in Tepic and Kalamazoo, methodologies and practices are carried out following community protocols and principles for the interaction and coexistence of people with Native seeds. In planting, we include consent practices such as the cultivation and planting ceremony to consult with the Elders about their opinion of the growth of the Milpa. The Elders consult, following their tradition, the ancestors to know how to proceed with the milpa. In Tepic, we cultivate with Wixárika families living in urban centers but with connections with rural communities in La Sierra. Tepic is in the state of Nayarit, Mexico. Most of the population is mestizo (mixed heritage) but many Indigenous communities continue living or migrating to the city. In Kalamazoo, we cultivate in the Hoop House one of the growing gardens at Kalamazoo College. Kalamazoo is in southwest Michigan.

The cultivation of the Milpa has been a school for learning not only about seeds (how to adapt to other climates) but also about ontological relationships with more-than-human beings. With the ceremonies and small practices that are made as an offering to the crop, we learn what it is like to be in a community with more than humans. The Milpa has become a university of the earth where with our bodily labors we enter communion with the seeds. It is also a project that connects us with the communities of Mexico and the USA. We estimated that the seeds will be shared with the community in Kalamazoo to be able to grow Milpas in various neighborhoods of Spanish-speaking peoples and some Indigenous communities in Michigan. This project will help migrants from Mexico and Central America to reconnect with ancestral seeds and ancestral practices.

4. The diversity of La Milpa

Native Corn offers a framework to understand human and more-than-human interactions by instructing us about how plants are teachers in a human's life. Settler colonialism and its effects disrupt the ways in which Native communities related to nonhumans. In Mexico, the commodification of Native species of Corn and other plants for human consumption and use in the pharmaceutical industry disrupted the "kincentric" relationship between humans and plants [22]. This kincentric relationship maintains Our Mother Corn as central to the life of families for not only ceremonial purposes but in the daily life of families to provide a framework to live life in wellness. Since 2021, students at Kalamazoo College have been cultivating Native

Corn seeds in the Hoop House. This action has been performed to maintain and sustain the livelihood of Native Corn from Mexico in the USA. However, cultivating Corn without a community contradicts the notions of kinship. For this, students at Kalamazoo College every year make a commitment to continue learning from Native seeds. In the summer of 2023, students from K learn from Our Mother Corn the multiple ways of growing in a different environment. Building on the work of the previous years, during the cultivation and the harvest, students measure and record the growth of the plants to assess how the local environment can help or support the growth of Our Mother Corn. The objective of the project is to predict how can we better adapt Native seeds from Mexico in Michigan. Furthermore, this project aims to understand the different changes in climate that can benefit or impact the growth of Corn. Students conduct research on soil, rain patterns, and temperatures to compare the results with the environment in Mexico. Finally, students will share with the community, their findings, and results to continue growing native seeds from Mexico.

In a study, Woznicki et al. [23] examine the effects of climate change on the cultivation of Maize and soybeans to understand the demands of irrigation in the Kalamazoo River Watershed of Michigan. The authors demonstrate that "there will likely be less water available during the growing season in the future, or evapotranspiration will be hindered due to temperatures stress in peak developments of corn" ([23], p. 252). This study suggests many adaptations to agricultural practices concerning irrigation to foresee future climate changes. Like this study, in the La Milpa Project we aim to predict some of the adaptations to maize in a new climate. Although Woznicki et al. do not discuss the type that maize cultivated; their conclusions are important to La Milpa Project. Due to the change in rain patterns in Kalamazoo, the project interns use an irrigation system in the garden. Another study by Schlüter et al. [24] discusses the different stress adaptions of maize such as low temperature, low nitrogen (n), and low phosphorus (P) stress. Like the study by Schlüter et al., the different types of stress that plants undergo in the Hoop House have an impact on the growth and the plant biomass [24]. However, many studies suggest that corn can be adapted to other climates using different agricultural strategies [25, 26]. Moradi et al. suggest to "consider the early maturing cultivars of maize in agro-ecosystems" ([26], p. 1236). In their study of maize adaptation in Iran, the authors selected "eight cultivars of maize with three growing stages periods" (1233: 2014). This selection allows for the collection of data and demonstrates the importance of early cultivation. On the other hand, Lorant et al. suggest that maize is genetically diverse and "worldwide consists of locally adapted open-pollinated populations (landraces) as well as modern inbred lines, derived from landraces, that are used in hybrid production for modern breeding" ([25], p. 676). All these studies take into consideration the different strategies for maize adaptation. The main goal of our study is to understand Maize adaptation to climate change.

4.1 Data collection

The first season was on May 10th, 2022. We cultivated five colors of Native Corn from Northern Mexico. We began the cultivation following traditional protocols and cultivated inside the Hoop House and outside in the rain garden. We started some seeds in planters to maximize germination. After five days, on May 15th, the seeds germinated and began the *Naika* ("sprout") stage, and we repotted them outside on May 29th and June 2nd. In the rain garden, we repotted the plants following the Wixárika arrangement of the Cornfield: in the front row, three plants of White Corn in the direction of the North, in the back two plants of Blue Corn in the direction of the south, to the right three Pink Corn in the direction of the West, to the left two Yellow Corn in the direction of the East, and two multi-color plants in the center. By June 23rd, our plants were in the vegetative stage or first leaf Y + ra ("Growing Greener"). The decision to cultivate the five variants of Our Mother Corn came as a community agreement. When I consulted with Felipa Rivera, she advised me to cultivate all the colors to have them together in the field. Along with Our Mother Corn, we cultivated flowers and other companion plants including tomatoes and kale inside the Hoop House, and milkweed and calendula in the rain garden. We used organic fertilizers, such as kelp fertilizer, to add extra potassium to the soil, and chicken manure for nitrogen, sulfur, and potassium. On July 12th, the plants reached a length of seven feet long inside the Hoop House and three feet long in the rain garden. On August 16th, the plants began the stage of tassel or *Tsakuluma, M + ayama, Tukima* ("To Tassel" or "VT Tassel"), and on August 27th, silking or K + paima. On August 30th, Yellow Corn was tasseled and on September 6th, the rest of the plants were tasseled. All the stages of Our Mother Corn corresponded with a Wixárika's personal names that connect human persons with plant persons [7].

Given the size of the growing garden, initially, Amy Newday (Mellon Fellow for Experiential Learning at Kalamazoo College) and myself were concerned with how close the population of plants was growing. With this, we began asking questions about cross-pollination. For many families in Tepic, cross-pollination is not really a concern. Families cultivate different variants of seeds and expect that some ears come with genetic modifications in color. Many families utilize this diverse Corn for corn-based drinks. In our first harvest, we were able to harvest a mature yellow ear of Our Mother Corn from the rain garden. In the Hoop House, there is an irrigation system and outside plants were watered with a water hose and rain. Inside the Hoop House, the plants grew over 2 m in length and matured very late in the season. On September 23rd of 2022, Nora Blanchard (Hoop House intern and K student) began hand-pollinating the plants. This was done with the hope of capturing pollen to help the plants in the process of pollination. This technique was useful because inside the Hoop House with the absence of wind, the pollen could not travel far. Due to the long days in Michigan, the plants were exposed to sunlight for about 12 h a day. Some plants released pollen early in the morning and some others released pollen in the afternoon (Data collected from personal conversations with Amy Newday and Nora Blanchard, Summer 2022).

On October 7th, the plants began the process of milking (*Saulima* or *Sutuli* stages). We prepared everything for the offerings during the harvest festival on campus. Some of the research questions we attempted to answer that summer were as follows: 1) What can we learn from genetics and cross-pollination? 2) How can the selection of seeds help to maximize the growth of plants in the next season? 3) Is it possible to cultivate multiple colors in a small population of plants? 4) How can we improve the fertility of plants? And 5) What can the roots teach us about Our Mother Corn?

The second cycle of Our Mother Corn began on May 20th of 2023. This time we decided to cultivate only yellow corn since based on the observations of the previous year these plants were more successful than the other colors. The seeds came from a selection of two batches (1) the seeds from Mexico and (2) the seeds from the fully mature ear of Corn from the previous year. Also, Zoe Reyes, the 2023 Hoop House intern, reported that the companion plants cultivated included: beans, squash, peas, amaranth, mullein, catnip, evening primrose, black-eyed Susan, daisy fleabane, and marigolds. The total number of Our Mother Corn plants cultivated was 48 plants in

12 mounts outside in the rain garden. We directly planted the seeds into the soil during our cultivation ceremony after the blessing of the seeds on May 23rd. In the following days, we visited Cornfield again for 5 days to follow the protocols of the Wixárika community and made some offerings. We did not see any seeds sprouting. A week after the cultivation on May 30th, some seeds sprouted. Also, Amy noticed that some seeds were missing, and others were not fully grown. The outside garden is the house of many other animals, insects, and plants. To avoid predators, Amy planted more seeds in pots inside the Hoop House and she transplanted those plants. On June 8th, some plants were transplanted. On June 13th, another batch of seeds was planted. Those seeds sprouted on June 22nd and then transplanted on June 27, 2023. The total population of plants was 45. Our research questions this season were: 1) how do the different seeding dates (May 23rd, June 7th, and June 13th) affect the flowering dates? 2) How can we improve the fertility of plants? What can we learn from our plants this season? 3) How is it growing in this environment affected by the soil, water, rain, and sun?

5. Challenges of planting native maize in Kalamazoo

Native Corn seeds are essential actors in the sustainable diet of many families in Mexico and Central America. The commitment and devotion of Wixárika families to *Tatéi Niwetsika* ("Our Mother Corn") and *Yuri'Ikú* ("True Corn") are essential for the survival of families and their genealogies, not only of human beings but beyond humans in El Gran Nayar. However, with the increase in families migrating to strategic points and other latitudes, families have been forced to find alternative ways to gain access to sustainable food in all geographies. In 2020, Mexico developed new laws that currently dictate the future of native seeds. Politicians approved a bill that gave farmers the right to grow landrace corn without the fear of growing near Genetically Modified Organism (GMO) fields. Corporations such as Monsanto have lost their power over the types of seeds to grow in Mexico. This development represents a giant leap in Mexican food forms, sovereignty, and agricultural sustainability. This favors native seeds since it is guaranteed that their growth outside of isolated communities can grow without being cross-pollinated and genetically damaged by hybrid seeds. According to Frabotta [27] and Peikes [28] Mexican President Andrés Manuel López Obrador (AMLO) decreed that by 2024, Mexico would eliminate the use of transgenic corn. This will positively influence the production and consumption of seeds. Soon, many Mexican farmers will be in need to change their paradigm and return to the traditional cultivation of seeds. To have transgenic-free agriculture means to adopt the traditional ways of planting crops to assure food security in the future.

Many Mexican citizens have realized the impact of GMOs and the consumption of GMOs in corn tortillas on the health of people and ecosystems. This helps families to make the decision of consuming Native seeds in a more conscious and responsible way. Although Maize was domesticated in Mesoamerica around 5500 years ago, today more than ever, both Indigenous and non-Indigenous communities advocate for the diversity of Native Maize in Mexico. This advocacy leads us to take up social and community commitment outside the context of local Native communities. Seeds, as well as people, also travel and seek other environments to germinate. Some seeds out of fear of people have lived in museum basements and out of the sunlight. Other seeds live guarded by the Elders with the fear of sowing and losing the last family seeds and their ancestral genealogy. Project on La Milpa in other latitudes helps us understand how seeds can be sown responsibly, ethically, and with awareness of social justice to grow plants that have crossed colonial borders. Thus, in this way, decolonize our diet and way of seeing corn as just another plant.

The contradictions and controversies about the La Milpa project have been various. During a recent panel presentation in El Gran Nayar, we had a dialog with the community in the audience. Some Elders shared their suggestions to cultivate earlier to avoid the cold from the north. Others expressed their concerns about the purpose of the project. A question that leaves us thinking and reflecting on the project is an ethical and moral question. The seeds of La Milpa Project in Tepic-Kalamazoo project belong to the Indigenous peoples of El Gran Nayar and they depend on the ceremony and ritual of many families. Within the Indigenous peoples, each family is clear about the role of each one in the care and distribution of the seeds. For example, in Y + rata, seeds are only given to those with spiritual responsibilities and community roles [7]. It is necessary to emphasize that many families feed themselves and support their families in many areas that include the cultural and ceremonial aspects of these seeds. Also, some families within the Wixárika community do not have enough space to plant each year and their seeds are saved for fear of losing them. In the La Milpa Project, we are very aware of the goals of the project. We aim to understand how Native Maize from Northern Mexico adapt to other climates with the hope of bridging communities together.

6. Conclusions

This chapter highlights the traditional techniques of cultivating Maize from ceremonial spaces to the Cornfield, with the intention of outlining the importance of community participation in the academic investigation. The author discusses the challenges of adapting to Native seeds in Michigan and the different perspectives of communities in Tepic and Kalamazoo in relationship to the growth of Our Mother Corn. The chapter emphasizes the importance of cultivating Native seeds employing traditional agriculture to maintain the agency of plants and learn from the seeds sustainable and reciprocal ways to relate with the land. The Milpa becomes a school that teaches us about diversity, responsibility, sacrifice, and physical and spiritual labor. The seeds teach us as agents of change the importance of sharing with other people and having responsibilities. The relationship is reciprocal since if people do not take care of the seeds, the seeds will give us fruits. Undoubtedly, people learn about caring for a sentient being, with agency, and effect on humans. In *La Milpa Project*, students learn that seeds are another type of being with the will and intention to germinate, grow, and bear fruit. When a plant does not grow, it does not have the will to teach us or learn from the new environment. They are simply not interested in growing, even if they are given care for their growth.

A key question in the work of food sovereignty is the question of whom the seeds belong to, specifically who owns or oversees the dispensation of the Native Corn seeds. In the La Milpa project, we acknowledge that the Indigenous peoples of Gran Nayar have developed scientific methods to guarantee a reciprocal relationship with Maize. For good germination, growth, and development of plants, specifically Native Corn seeds, a community or a group of people must follow an ontological relationship with seeds. When that relationship is lost, the way we see other beings is also lost. The ontological relationship between people and the land are key concepts in the philosophy of Indigenous peoples. Handmade tortillas made with Native Corn still preserve the reciprocal relationship between people and plants. People make tortillas and other Corn-based foods as offerings during cultivation. With a change in ideology, the commodification of Corn-based foods began in Mexico and abroad. In the La Milpa project, we acknowledge the importance of the connection between seeds and community, specifically speakers of the Wixárika language. For students, they learn about specific methodologies and epistemologies to conceive plants like Our Mother Corn as kin. The different components of the project are necessary for social change, not only to reverse colonial ideas to conceive food but to decolonize our diet and find ways to secure food in times of climate change and food scarcity.

Acknowledgements

I acknowledge that part of this chapter was written in the unceded land of the Council of the Three Fires—the Ojibwe, the Odawa, and the Potawatomi. Additionally, I wrote this chapter in El Gran Nayar, Náayeri, O'dam, Mexikan, and Wixárika unceded homelands. I thank my teachers, plant, human, and more-than-human relatives for accompanying me through life. Especially to Our Mother Corn for teaching me about kinship. Also, I am grateful to my *comadre* Felipa Rivera Lemus, and my godmother Rosalía Lemus de la Rosa for helping me collect oral traditions and teachings from women in Y + rata community. I am thankful for the support of Alison Geist, Amy Newday, Sara Stockwood, and Margaret Wiedenhoeft from Kalamazoo College. This work could not have been possible without the support of the Mary Jane Underwood Stryker Institute for Service-Learning, Center for Environmental Stewardship, and Center for International Programs. Finally, much gratitude to Zoe Reyes and Nora Blanchard for their work during their internships in the Hoop House. All the comments and opinions are my responsibility.

Author details

Cyndy Garcia-Weyandt
Kalamazoo College, Kalamazoo, Michigan, USA

*Address all correspondence to: cyndy.garcia-weyandt@kzoo.edu

References

[1] Staller J. Maize Cobs and Cultures: History of *Zea mays* L. New York: Springer Science & Business Media; 2009

[2] Piperno DR, Flannery KV. The earliest archaeological maize (*Zea mays* L.) from highland Mexico: New accelerator mass spectrometry dates and their implications. Proceedings of the National Academy of Sciences of the United States of America. 2001;**98**:2101-2103. DOI: 10.1073/pnas.98.4.2101

[3] Cruikshank J. Glaciers and climate change: Perspectives from oral tradition. Arctic. 2001;**1**:377-393

[4] Menzies CR, editor. Traditional Ecological Knowledge and Natural Resource Management. Lincoln, NE: University of Nebraska Press; 2006

[5] Nadasdy P. The politics of TEK: Power and the "integration" of knowledge. Arctic Anthropology. 1999;**1**:1-8

[6] Lemus FR, Zing R. In: Fikes JC, Weigand PC, de Weigand CG, editors. Huichol Mythology. Tucson: University of Arizona Press; 2004

[7] Garcia-Weyandt CM. Te 'uayemat+ ta Kiekari Tatéi Niwetsikak+: Urban Wixárika Healing Practices and Ontology (Diss). Los Angeles: University of California; 2020

[8] Vidal Martínez VA. Proyecto FZ002: Conocimiento de la diversidad y distribución actual de maíz nativo y sus parientes silvestres de México. In: Comisión Nacional para el Conocimiento y Uso de la Biodiversidad (CONABIO). Tepic, Nayarit, Septiembre del 2008

[9] Broda J. Cosmovisión Mesoamericana y ritualidad agrícola: Estudios interdisciplinarios y regionales. Puebla: Benemérita Universidad Autónoma de Puebla, Vicerrectorado de Docencia, Dirección de Fomento Editiorial; 2009

[10] Burgos-Ellis H. The sexual life of plants in Mesoamerica. Natural Things in Early Modern Worlds. 2023

[11] Furst PT. The Parching of the Maize: An Essay on the Survival of Huichol Ritual. Vol. 3. E. Stiglmayr. Harvard; 1968

[12] López Moreno I, Bordi IV. El maíz nativo en México. Una aproximación crítica desde los estudios rurales. Juan Pablos: Universidad Autónoma Metropolitana; 2016

[13] Morton PE. Tortillas: A Cultural History. Albuquerque, NM: UNM Press; 2014

[14] Ruvalcaba J. Vida cotidiana y consumo de maíz en la huasteca veracruzana. México: Centro de Investigaciones y Estudios Superiores en Antropología Social; 1987

[15] Sagastume RC. El costumbre y la soberanía alimentaria Wixárika. Guadalajara, Jalisco: Universidad de Guadalajara; 2010

[16] Kremen C, Miles A. Ecosystem services in biologically diversified versus conventional farming systems: Benefits, externalities, and trade-offs. Ecology and Society. 2012;**17**(4). Available from: http://www.jstor.org/stable/26269237

[17] Garcia-Weyandt CM. Curing with Our Mother Corn. The Jugaad Project. 2021. Available from: www.thejugaadproject.pub/home/mother-corn

[18] Kimmerer R. Braiding Sweetgrass: Indigenous Wisdom, Scientific Knowledge, and the Teachings of Plants. Canada: Milkweed Editions; 2013

[19] Santillán Santillán M, Chimba Samba LF. Ishkay Yachay: Propuesta de Educación Intercultural Bilingüe para vigorizar los Saberes Ancestrales Andinos en equivalencia con la Modernidad. Ecuador: Editorial Yachay Wasi; 2016. Available from: http://8.242.217.84:8080/jspui/handle/123456789/33353

[20] Atalay S, McCleary AC, editors. The Community-Based PhD: Complexities and Triumphs of Conducting CBPR. Tucson, AZ: University of Arizona Press; 2022

[21] Simpson LB. Land as pedagogy: Nishnaabeg intelligence and rebellious transformation. Decolonization: Indigeneity, Education & Society. 2014;**3**(3):1-24

[22] Salmón E. Eating the Landscape: American Indian Stories of Food, Identity, and Resilience. Tucson, AZ: University of Arizona Press; 2012

[23] Woznicki SA, Pouyan NA, Parsinejad M. Climate change and irrigation demand: Uncertainty and adaptation. Journal of Hydrology: Regional Studies. 2015;**3**:247-264. DOI: 10.1016/j.ejrh.2014.12.003

[24] Schlüter U, Colmsee C, Scholz U, Bräutigam A, Weber AP, Zellerhoff N, et al. Adaptation of maize source leaf metabolism to stress related disturbances in carbon, nitrogen and phosphorus balance. BMC Genomics. 2013;**14**:1-25. DOI: 10.1186/1471-2164-14-442

[25] Lorant A, Ross-Ibarra J, Tenaillon M. Genomics of long- and short-term adaptation in maize and teosintes. In: Dutheil JY, editor. Statistical Population Genomics. Methods in Molecular Biology. Vol. 2090. New York: Humana; 2020. DOI: 10.1007/978-1-0716-0199-0_12

[26] Moradi R, Koocheki A, Nassiri MM. Adaptation of maize to climate change impacts in Iran. Mitigation and Adaptation Strategies for Global Change. 2014;**19**:1223-1238

[27] Frabotta D. Mexico court upholds presidential decree to ban glyphosate, GMOs. AgriBusiness Global. 2021. Available from: https://www.agribusinessglobal.com/markets/mexico-court-upholds-presidential-decree-to-ban-glyphosate-gmos/

[28] Peikes K. Mexico plans to ban all GMO corn. Now a key deadline looms as the U.S. pushes back. KCUR. 2023. Available from: https://www.kcur.org/2023-03-29/mexico-plans-to-ban-all-gmo-corn-now-a-key-deadline-looms-as-the-u-s-pushes-back

Section 2

Nutrient Planning and Control of Maize

Chapter 3

Nutrient Management of Maize

Maryam Batool

Abstract

This chapter presents a comprehensive overview of nutrient management practices tailored for optimizing maize production. It covers critical aspects, including soil testing protocols, advanced fertilizer application methods, organic and inorganic amendments, precision nutrient management approaches, integrated strategies, and conservation agriculture-based practices. Recognizing maize's significance for global food security and economic prosperity, the chapter emphasizes efficient and sustainable nutrient management to achieve high yields. Precision technologies enable targeted fertilizer applications, while organic and inorganic amendments enhance soil fertility and nutrient cycling. Integrated nutrient management reduces environmental risks and improves long-term soil fertility. Conservation agriculture-based practices, such as reduced tillage and cover cropping, positively influence maize yield and sustainability by enhancing nutrient retention and water management. Overall, adopting appropriate nutrient management practices is crucial for maximizing maize production while ensuring food security and environmental well-being.

Keywords: maize, nutrient management, soil testing, fertilizers, organic amendments, micronutrients, balanced nutrition, yield, environment

1. Introduction

Maize (*Zea mays* L.) is an important cereal crop globally, with a production of over 1.1 billion metric tons in 2020 [1]. Maize is not only a staple food for humans but also an essential feed for livestock. In addition, it is widely used for the production of biofuels and industrial products. Therefore, increasing maize productivity is crucial for food and nutritional security, as well as for economic growth. Nutrient management is one of the critical factors that can significantly affect maize yield and quality. Proper nutrient management not only ensures an adequate supply of essential nutrients but also improves soil health and reduces environmental pollution. The nutrient management practices for maize production can vary depending on soil fertility, climatic conditions, crop management practices, and other factors. In recent years, several studies have focused on improving nutrient management practices for maize production. These studies have explored various approaches such as balanced fertilization, precision nutrient management, integrated nutrient management, and the use of organic and inorganic amendments to improve soil fertility and nutrient use efficiency [2–5]. Studies have shown that conservation agriculture-based practices such as zero-till flatbed (ZTFB) and permanent beds (PNB) can produce greater maize grain-equivalent yield (MGEY) compared to conventional tillage (CT). Similarly, nutrient expert-based application (NE) and recommended fertilization (RDF) have been found to increase MGEY compared to farmers' fertilizer practices (FFP). Furthermore, these practices have been shown to enhance soil properties, including bulk density and microbial biomass carbon. Integrated nutrient management involves customizing nutrient use by considering contributions through residue retention, atmospheric nitrogen fixation, and residual nutrients. This approach has been shown to enhance maize yield, nutrient uptake, and economic returns, compared to using only organic or inorganic fertilizers. These approaches have shown promising results in enhancing maize yield, reducing input costs, and promoting sustainable agricultural practices [6–10]. This chapter aims to review the latest findings on nutrient management practices for maize production and their implications for sustainable agriculture. The chapter will cover various aspects of nutrient management, including soil testing, fertilizer application methods, timing, and rates, as well as the use of organic and inorganic amendments. The chapter will also discuss the challenges and opportunities for improving nutrient management practices in maize production. Overall, this chapter intends to provide insights into how nutrient management practices can contribute to sustainable maize production and food security.

2. The importance of maize production

Maize, also known as corn (*Zea mays* L.), is a highly versatile crop with a long history of domestication dating back 9000 years ago. Its global production has been increasing rapidly over the past few decades due to technological advancements, yield improvements, and area expansion driven by rising demand. Maize has become the most widely grown and traded crop and is currently the leading cereal in terms of production volume [11]. Maize plays a crucial role in global agri-food systems as a multi-purpose crop. It is primarily used as a feed for livestock, but it is also an important food crop, especially in sub-Saharan Africa and Latin America, where it serves as a staple food for millions of people [12]. Additionally, maize is used in many non-food products such as biofuels, starches, and sweeteners [13]. Maize production has the potential to address several pressing global challenges, including food and nutritional security, water scarcity, and climate change. In sub-Saharan Africa, maize is an essential crop for smallholder farmers and provides a vital source of income and food. It is estimated that maize is cultivated on over 33 million hectares of land in sub-Saharan Africa, with over 300 million people relying on it as a source of food and income [14]. Maize is also a crop with wide adaptability under different ecological scenarios, making it an essential crop for sustainable agriculture. In India, the conventional rice-wheat cropping system (RWCS) has been the dominant production system in the Indo-Gangetic Plains (IGPs). However, this cropping system has faced several sustainability challenges due to the high water requirement, soil fertility degradation, and inefficient input usage [15]. To address these challenges, conservation agriculture (CA) practices based on maize production have been introduced to enhance resource use efficiency, restore soil fertility, and improve crop yields. Maize production is critical for global food and nutritional security, with its versatile uses making it a vital component of the agri-food system. Maize also has the potential to address sustainability challenges, such as water scarcity and climate change, making it a crucial crop for sustainable agriculture.

2.1 Nutrient management basics

Nutrient management involves managing the amount, source, placement, form, and timing of the application of plant nutrients and soil amendments to optimize plant growth and yield while minimizing environmental impact. Integrated nutrient management (INM) is a recommended practice that involves using both organic and inorganic fertilizers to improve soil productivity and crop productivity. This approach, along with the integrated use of major plant nutrients (nitrogen, phosphorus, and potash), organic carbon sources (animal manures and plant residues), and bio-fertilizers (beneficial microbes), has been shown to significantly enhance maize growth, yield, and yield components, as well as grower's income. Conservation agriculture (CA) practices, including zero-till flatbed (ZTFB), permanent beds (PNB), and conventional systems (CT), have also been found to increase farm profits and improve soil properties. Nutrient expert-based application (NE), recommended fertilization (RDF), and farmers' fertilizer practice (FFP) are recommended CA-based nutrient management practices that can further enhance productivity and profitability [2, 6, 16]. Maize production heavily relies on adequate nutrient management, with nitrogen, phosphorus, and potassium being the most critical nutrients. Nitrogen is vital for vegetative growth and grain yield, but its mismanagement can cause environmental problems such as nitrate leaching and greenhouse gas emissions. Various nitrogen management practices, including split applications during planting and vegetative stages, have been found effective in improving maize yields and nitrogen use efficiency. Similarly, phosphorus plays a critical role in root growth, flowering, and grain filling, and its deficiency can result in poor crop quality and reduced yield [17–19]. Phosphorus management practices, such as soil testing and banding phosphorus fertilizers, have been found to enhance phosphorus availability in the soil and improve maize productivity. Additionally, potassium is essential for osmoregulation, enzyme activation, and photosynthesis, and its deficiency can lead to reduced yield and increased susceptibility to biotic and abiotic stresses. Effective potassium management practices include soil testing, potassium fertilizer application, and applying potassium fertilizer at planting and during the vegetative stage. Research has shown that these practices can improve maize yield and potassium use efficiency [4, 20, 21]. Understanding the nutrient requirements of maize, as well as the nutrient content of the soil, is essential to develop a nutrient management plan that balances these needs with available resources.

2.2 Soil testing for maize production

Soil testing holds a pivotal role in optimizing nutrient management specifically tailored for maize production. By analyzing soil samples, farmers gain invaluable insights into the soil's nutrient content and pH levels, enabling them to make well-informed decisions regarding fertilizer application. Recent research papers have extensively highlighted the profound significance of soil testing in this context. In a notable study conducted between 2015 and 2016, the focus was on bridging the maize yield gap and enhancing soil properties in coastal saline soil. The researchers explored the efficacy of a combined application of flue gas desulfurization gypsum and furfural residue (known as CA). Intriguingly, the post-harvest CA treatment exhibited remarkable outcomes, with notable increases observed in calcium (Ca2+) and soil organic carbon (SOC) contents, while simultaneously reducing sodium (Na+) content and pH levels in the upper soil layer. Consequently, maize crops experienced significant enhancements in nitrogen, phosphorus, potassium, calcium, and magnesium accumulations, alongside a decrease in Na accumulation when compared to the control group [22]. Another noteworthy study delved into the dynamics of global maize production, consumption, and trade, aiming to decipher evolving trends over the past 25 years and their consequential impact on research and development (R&D), with a particular focus on the Global South. The study emphasized the pressing need for augmented investments in R&D endeavors to fortify maize's pivotal role in ensuring food security, sustaining livelihoods and effectively intensifying production, all while adhering to the constraints imposed by planetary boundaries [23]. These research findings substantiate the indispensability of soil testing in the realm of maize production. Moreover, they underscore the necessity for further exploration to develop innovative and more potent methodologies aimed at improving soil properties and elevating maize yields. As such, these insights reinforce the critical role that soil testing plays in optimizing nutrient management strategies, customizing fertilizer application practices, and addressing the overarching global challenges associated with maize cultivation. Soil testing occupies a central position in the intricate web of nutrient management for maize production. Recent research profoundly accentuates its significance in fine-tuning fertilizer application, ameliorating soil characteristics, and ultimately bolstering maize yields. By diligently incorporating soil testing into their agricultural practices, farmers can aptly discern the most optimal courses of action, thereby maximizing nutrient utilization, mitigating environmental repercussions, and fostering sustainable and prosperous maize farming [24–27].

2.3 Fertilizer types and application methods

Nutrient management plays a vital role in optimizing maize production and selecting appropriate fertilizer types and application methods is crucial for achieving optimal crop yields [28]. Maize requires specific nutrients, including nitrogen (N), phosphorus (P), and potassium (K), as well as secondary and micronutrients, to support its growth and development. Nitrogen fertilizers, such as urea, ammonium nitrate, and ammonium sulfate, are commonly used to supply the essential nutrient nitrogen to maize. Nitrogen application should be split into multiple doses to match the crop's demand throughout the growing season [29]. Phosphorus fertilizers, such as diammonium phosphate (DAP) and triple superphosphate (TSP), are beneficial for root development and overall plant growth. These fertilizers are typically applied at planting time, either broadcast or as a band near the seed, to ensure efficient uptake by the developing root system. Potassium fertilizers, such as potassium chloride (KCl) and potassium sulfate (K2SO4), are crucial for enhancing maize yield and improving drought tolerance [30]. The application of potassium can be incorporated into the soil before planting or applied as a side dress during the early stages of crop growth. Additionally, secondary nutrients like calcium (Ca), magnesium (Mg), and sulfur (S), along with micronutrients like zinc (Zn), copper (Cu), iron (Fe), manganese (Mn), and boron (B), and molybdenum (Mo), play significant roles in maize production [31]. These nutrients can be supplied through soil amendments or foliar applications, based on soil test results and crop nutrient requirements. Appropriate fertilizer application methods, such as broadcasting, banding, side-dressing, and foliar spraying, should be employed to ensure efficient nutrient uptake and minimize losses. By following recommended nutrient management practices, including split applications and considering the specific nutrient requirements of maize, farmers can achieve higher yields and sustainable crop production [31–33].

2.4 Timing and rates of fertilizer application

Timing and rates of fertilizer application are crucial factors in optimizing maize production and ensuring efficient nutrient uptake. Nitrogen (N) is a key nutrient for maize, and it should be applied in multiple doses to meet the crop's demand throughout the growing season [33]. The first application of nitrogen can be done at planting time, with subsequent doses applied during the early vegetative stage and at the onset of the rapid growth phase [34]. Phosphorus (P) is essential for root development and overall plant growth. It is recommended to apply phosphorus-based fertilizers, such as diammonium phosphate (DAP) or triple superphosphate (TSP), at planting time either as a broadcast or band application near the seed [35]. The application of potassium (K) is beneficial for enhancing maize yield and improving drought tolerance. Potassium fertilizers like potassium chloride (KCl) or potassium sulfate (K2SO4) can be incorporated into the soil before planting or applied as a side-dress during the early growth stages [32]. Additionally, secondary nutrients such as calcium (Ca), magnesium (Mg), and sulfur (S), along with micronutrients including zinc (Zn), copper (Cu), iron (Fe), manganese (Mn), boron (B), and molybdenum (Mo), are important for maize production. The application rates of these nutrients depend on soil test results and crop nutrient requirements [36]. Generally, it is recommended to follow regional fertilizer recommendation guidelines to determine the appropriate rates of nutrient application for maize [37]. By carefully timing and applying fertilizers at the right rates, farmers can ensure an adequate nutrient supply for maize and maximize crop productivity.

2.5 Organic and inorganic amendments

In maize nutrient management, the use of organic and inorganic amendments plays a crucial role in improving soil fertility and providing essential nutrients for optimal crop growth [33]. Organic amendments, such as farmyard manure (FYM), compost, and green manure, are valuable sources of organic matter and nutrients [38]. These amendments enhance soil structure, water-holding capacity, and nutrient availability, thereby promoting maize growth and productivity. Incorporating organic amendments into the soil before planting or as a top dressing during the growing season can effectively supply nutrients like nitrogen, phosphorus, and potassium [39]. In addition to organic amendments, inorganic fertilizers are widely used to supplement nutrient requirements in maize production. Nitrogen-based fertilizers, such as urea, ammonium nitrate, and ammonium sulfate, provide readily available nitrogen for optimal plant growth [40–42]. Phosphorus fertilizers, such as diammonium phosphate (DAP) and triple superphosphate (TSP), are important for promoting root development and enhancing yield potential. Potassium fertilizers, including potassium chloride (KCl) and potassium sulfate (K2SO4), are essential for improving maize yield and stress tolerance [43]. Applying inorganic fertilizers in a targeted manner, such as banding or side-dressing, can maximize nutrient uptake and minimize losses. The combination of organic and inorganic amendments in maize nutrient management can optimize nutrient availability, improve soil fertility, and support sustainable crop production [44]. It is important to consider the nutrient requirements of maize, soil nutrient levels, and local agricultural practices when determining the appropriate application rates and timing of organic and inorganic amendments. By implementing effective nutrient management strategies using a combination of organic and inorganic amendments, farmers can enhance maize productivity while minimizing environmental impacts [45].

2.6 Precision nutrient management

Precision nutrient management for maize plays a pivotal role in optimizing crop productivity while minimizing environmental impacts associated with excessive fertilizer use. Precision nutrient management refers to the precise application of fertilizers based on the specific nutrient needs of the crop, considering factors such as soil variability, crop growth stage, and yield potential [46]. This approach involves utilizing advanced technologies such as remote sensing, geographic information systems (GIS), and variable rate application (VRA) systems to spatially and temporally tailor nutrient application rates. Precision nutrient management helps to optimize fertilizer use efficiency and reduce nutrient losses through targeted application, thus improving crop performance and minimizing environmental risks [47]. Remote sensing technologies, including satellite imagery and aerial drones, provide valuable information about crop health and nutrient status. These technologies enable the identification of nutrient deficiencies or excesses in specific areas of the field, allowing farmers to apply fertilizers precisely where they are needed [48]. GIS-based soil mapping and soil nutrient testing further assist in identifying nutrient variability across the field, enabling site-specific nutrient recommendations. Variable rate application systems enable farmers to apply fertilizers at different rates within a field, based on site-specific recommendations. By adjusting fertilizer rates based on the variability of soil nutrient levels, farmers can ensure that nutrients are provided in optimal quantities, maximizing crop uptake and minimizing losses. This approach also helps to avoid the over-application of nutrients in areas where they are not needed, reducing the risk of nutrient runoff into water bodies. Generally, precision nutrient management for maize offers a sustainable and efficient approach to fertilizer application. By utilizing advanced technologies and tailoring nutrient application rates to the specific needs of the crop and field, farmers can achieve higher yields, reduce fertilizer costs, and minimize environmental impacts associated with nutrient losses. Implementing precision nutrient management practices can contribute to the long-term sustainability and profitability of maize production systems [20, 49, 50].

2.7 Integrated nutrient management

Integrated nutrient management (INM) plays a crucial role in optimizing maize production by adopting a holistic approach to meet crop nutrient requirements efficiently [2, 3]. INM involves the integration of various nutrient sources, including organic manures, inorganic fertilizers, biofertilizers, and crop residues, to enhance soil fertility and promote sustainable crop growth. Organic manures, such as farmyard manure (FYM) and compost, are valuable sources of macro and micronutrients, as well as organic matter, which improve soil structure and nutrient availability. Incorporating organic manures into the soil at recommended rates not only supplies essential nutrients but also enhances soil health and microbial activity. Inorganic fertilizers, such as nitrogen, phosphorus, and potassium fertilizers, are often used in combination with organic manures to supplement nutrient deficiencies and achieve balanced nutrition. Biofertilizers, including nitrogen-fixing bacteria, phosphate-solubilizing bacteria, and mycorrhizal fungi, can be applied either as seed inoculants or through soil application to enhance nutrient uptake and improve soil nutrient cycling [51]. Additionally, incorporating crop residues into the soil as green manure helps enhance soil organic matter content and nutrient availability. INM practices also include precision nutrient management based on soil testing to determine nutrient deficiencies and adjust fertilizer application rates accordingly. Adopting balanced fertilization practices through INM not only ensures optimal nutrient supply to maize but also promotes environmental sustainability by minimizing nutrient losses and reducing the risk of pollution [5, 38, 42]. By integrating organic manures, inorganic fertilizers, biofertilizers, and crop residues, along with precision nutrient management, farmers can achieve improved maize productivity and maintain soil fertility in a sustainable manner. While SSNM (site-specific nutrient management) is able to be tailored to the requirements of a site or field, for a broader purpose, INM provides better nutrient management [52].

2.8 Conservation agriculture-based practices

Conservation agriculture (CA) is an approach that promotes sustainable and environmentally friendly maize production while enhancing soil health and crop resilience [7, 15]. Several CA-based practices have proven effective in maize cultivation. One key practice is minimum soil disturbance, which involves reducing or eliminating conventional tillage to preserve soil structure and prevent erosion [53]. Zero tillage, where seeds are directly planted into untilled soil, has shown positive effects on maize yields by improving water infiltration and conserving soil moisture [6, 7]. Another important practice is residue management, where crop residues are left on the soil surface instead of being removed or burned. This practice enhances organic matter content, improves soil fertility, and reduces weed pressure [15]. Cover cropping is also integral to CA in maize systems, where non-commercial crops are grown during fallow periods to protect the soil from erosion, suppress weeds, and improve nutrient cycling [54]. Additionally, crop rotation is a key component of CA, as it breaks disease and pest cycles, improves soil structure, and enhances nutrient availability [55]. Intercropping, the simultaneous cultivation of two or more crops in close proximity, is another beneficial CA practice that optimizes resource use and diversifies farm income [56]. Precision nutrient management, including site-specific fertilization based on soil testing and variable rate application, helps optimize nutrient use efficiency while minimizing environmental impacts. Effective weed management through integrated approaches, such as using cover crops, mechanical methods, and targeted herbicide application, is essential in CA maize production to reduce weed competition. By adopting these CA-based practices, maize producers can achieve sustainable crop production, improve soil health, and mitigate environmental risks [57, 58].

2.9 Best practices for nutrient management in maize production

Implementing best practices for nutrient management is crucial for optimizing maize production and ensuring sustainable crop yields. Firstly, conducting regular soil testing is essential to assess nutrient levels and pH, providing valuable information for fertilizer recommendations [5, 6]. Splitting nitrogen (N) applications throughout the growing season based on crop demand is highly recommended to improve nitrogen use efficiency [25]. For phosphorus (P) fertilization, applying diammonium phosphate (DAP) or triple superphosphate (TSP) at planting time, either broadcast or as a band near the seed, promotes root development and overall plant growth [20]. Potassium (K) fertilizers should be applied either as a pre-plant incorporation or as a side-dress during early crop stages to enhance maize yield and improve drought tolerance. In addition to N, P, and K, secondary nutrients (calcium, magnesium, and sulfur) and micronutrients (zinc, copper, iron, manganese, boron, and molybdenum) play vital roles in maize production. Soil amendments or foliar applications can be utilized to address deficiencies based on soil test results and crop nutrient requirements [49]. Employing appropriate fertilizer application methods such as broadcasting, banding, side-dressing, or foliar spraying ensures efficient nutrient uptake and minimizes losses [38]. Moreover, adopting conservation practices such as cover cropping, crop rotation, and precision farming techniques can improve nutrient cycling, reduce nutrient runoff, and enhance soil fertility. Multiple studies have linked the impact of biochar on crop productivity to various factors, including enhanced cation exchange capacity (CEC) and the subsequent retention of nutrients, elevated pH levels and increased base saturation, augmented availability of phosphorus, and improved water accessibility for plants [59]. Regular monitoring of crop health and adjusting fertilizer applications based on visual symptoms or plant tissue analysis is crucial to avoid over or under-application of nutrients. By adhering to these best practices, farmers can optimize nutrient management in maize production, leading to increased yields, improved resource use efficiency, and environmental sustainability [60].

2.10 Challenges and opportunities for improving nutrient management practices

Effective nutrient management is essential for sustainable agriculture and maximizing crop productivity, but it faces several challenges and offers opportunities for improvement. One major challenge is the improper use of fertilizers, resulting in nutrient imbalances, environmental pollution, and economic losses [61]. Over-application of fertilizers can lead to nutrient runoff, causing water pollution and eutrophication [62]. On the other hand, insufficient fertilizer application can result in nutrient deficiencies, limiting crop yields. Another challenge is the lack of soil testing and nutrient analysis, which hinders precise fertilizer recommendations based on the specific nutrient requirements of crops. Inadequate knowledge and awareness among farmers regarding nutrient management practices further contribute to suboptimal fertilizer use [16, 31, 39]. However, there are opportunities for enhancing nutrient management practices. The development and promotion of precision agriculture technologies enable site-specific nutrient application, optimizing fertilizer use efficiency [63]. Integration of organic farming practices, such as cover cropping, crop rotation, and the use of organic amendments, can enhance soil fertility and reduce the reliance on synthetic fertilizers [64]. Additionally, implementing conservation practices like conservation tillage and nutrient management planning can minimize nutrient losses and improve nutrient use efficiency [50]. Education and extension programs play a crucial role in increasing farmers' understanding of nutrient management principles and practices, encouraging adoption of sustainable approaches. Furthermore, research efforts are focused on developing advanced fertilizers with slow-release mechanisms and improved nutrient uptake efficiency. By addressing these challenges and embracing the opportunities, sustainable nutrient management practices can be achieved, promoting environmentally friendly agriculture and ensuring long-term food security [65, 66].

3. Conclusion

In conclusion, effective management of fertilizer types and application methods is crucial for maximizing maize productivity. Nitrogen, phosphorus, potassium, secondary nutrients, and micronutrients play vital roles in supporting the growth and development of maize plants. Splitting nitrogen applications throughout the growing season to match crop demand and using fertilizers like urea, ammonium nitrate, and ammonium sulfate can ensure optimal nitrogen supply. Phosphorus fertilizers such as diammonium phosphate and triple superphosphate are beneficial for root development and should be applied at planting time either through broadcasting or banding near the seed. Potassium fertilizers like potassium chloride and potassium sulfate can enhance maize yield and improve drought tolerance and should be applied before planting or as a side-dress during early crop growth. Additionally, the application of secondary nutrients and micronutrients, based on soil test results and crop requirements, can significantly contribute to maize production. It is important to consider appropriate fertilizer application methods such as broadcasting, banding, side-dressing, and foliar spraying to ensure efficient nutrient uptake and minimize losses. Farmers should follow recommended nutrient management practices and tailor their fertilization strategies to meet the specific needs of their maize crops, as this will lead to higher yields and sustainable crop production. Regular soil testing and monitoring can provide valuable insights for adjusting fertilizer types and application methods to optimize maize nutrient management and improve overall productivity.

Nomenclature

N	nitrogen
P	phosphorus
K	potassium
Ca	calcium
Mg	magnesium
S	sulfur
Fe	iron
Mn	manganese
Zn	zinc
Cu	copper
B	boron
Mo	molybdenum
pH	soil pH level
OM	organic matter
EC	electrical conductivity
ANR	annual nutrient requirement
TSP	triple superphosphate
DAP	di-ammonium phosphate
MAP	mono-ammonium phosphate
NPK	nitrogen-phosphorus-potassium
CAN	calcium ammonium nitrate
UAN	urea ammonium nitrate solution
FYM	farmyard manure
DCT	deep placement of fertilizer
SSNM	site-specific nutrient management
FFD	full fertilizer dose
LCC	leaf color chart
CF	crop factor
ICM	integrated crop management

VRA	variable rate application
DSS	decision support system
RZWQM	root zone water quality model
UAV	unmanned aerial vehicle
GIS	geographic information system
RUSLE	revised universal soil loss equation
NUE	nutrient use efficiency
EONR	economic optimum nitrogen rate
BMP	best management practices
NIRS	near-infrared reflectance spectroscopy
PSNT	pre-sidedress soil nitrate test
RTK	real-time kinematic

Author details

Maryam Batool
COMSATS University, Islamabad, Pakistan

*Address all correspondence to: maryam.batool1416@gmail.com

References

[1] Food and Agriculture Organization of the United Nations (FAO). FAOSTAT Database. 2021. Available from: http://www.fao.org/faostat/en/#data/QC

[2] Gezahegn AM. Role of integrated nutrient management for sustainable maize production. International Journal of Agronomy. 27 May 2021;**2021**:1-7

[3] Kalhapure AH, Shete BT, Dhonde MB. Integrated nutrient management in maize (Zea mays L.) for increasing production with sustainability. International Journal of Agriculture Food Science & Technology. 2013;**4**:195-206

[4] Mohapatro S, Shankar T, Swami GVNS, Sahu S. A review on precision nutrient management in maize. Agro Economist-An International Journal. 2020;7(2):77-83

[5] Mahmood F, Khan I, Ashraf U, Shahzad T, Hussain S, Shahid M, et al. Effects of organic and inorganic manures on maize and their residual impact on soil physico-chemical properties. Journal of Soil Science and Plant Nutrition (Temuco, Chile). 2017;**17**(1). DOI: 10.4067/S0718-95162017005000002

[6] Pooniya V, Zhiipao RR, Biswakarma N, Jat SL, Kumar D, Parihar CM, et al. Long-term conservation agriculture and best nutrient management improves productivity and profitability coupled with soil properties of a maize–chickpea rotation. Scientific Reports. 2021;**11**:10386

[7] Pooniya V, Biswakarma N, Parihar CM, Swarnalakshmi K, Lama A, Zhiipao RR, et al. Six years of conservation agriculture and nutrient management in maize–mustard rotation: Impact on soil properties, system productivity and profitability. Field Crops Research. 2021;**260**:108002. DOI: 10.1016/j.fcr.2020.108002

[8] Patra D, Dawson J. Effect of nutrient management and intercropping with cowpea on growth and yield of maize (*Zea mays* L.). The Pharma Innovation. 2022;**11**(4):1105-1110

[9] Halleron R. Nutritional Requirements of Maize Crops [Internet]. Agriland; 2022. Available from: https://www.agriland.ie/farming-news/nutritional-requirements-of-maize-crops/#:~:text=Like%20many%20other%20crops%2C%20maize,it%27s%20importance%20for%20root%20growth [Accessed: May 18, 2023]

[10] Goud RP, Kumar KA, Srinivas A, Rao AM, Ramprakash T. Effect of integrated nutrient management on soil properties and fertility in baby corn (Zea mays. L)–Hyacinth bean (Lablab purpureus var. typicus) cropping system. The Pharma Innovation. 2023;**12**(3):5467-5473

[11] Chaudhary D, Kumar S, Yadav OP. Maize: Nutrition dynamics and novel uses. Research Gate. 2014. DOI: 10.1007/978-81-322-1623-0

[12] Erenstein O, Jaleta M, Sonder K, Mottaleb K, Prasanna BM. Global maize production, consumption and trade: Trends and R&D implications. Food Security. Oct 2022;**14**(5):1295-1319

[13] The Global Maize Market. Food and Agriculture Organization of the United Nations [Internet]. 2021. Available from: https://www.fao.org/markets-and-trade/publications/detail/en/c/1415511/ [Accessed: May 18, 2023]

[14] Jabran K, Ata Z, Farooq M. Maize: Cereal with a Variety of Uses [Internet].

Dawn; 2007. Available from: https://www.dawn.com/news/236949/maize-cereal-with-a-variety-of-uses [Accessed: May 18, 2023]

[15] Shrestha J, Subedi S, Timsina KP, Chaudhary A, Kandel M, Tripathi S. Conservation agriculture as an approach towards sustainable crop production: A review. Farming and Management. 2020;**5**(1):7-15

[16] Nisar S, Mavi MS, Singh J, Dey P. Integrated nutrient management in spring-maize improves yield, nutrient use efficiency and minimizes greenhouse gas intensity. Archives of Agronomy and Soil Science. 2023;**69**(13):2522-2536

[17] Cui Z, Chen X, Zhang F. Current nitrogen management status and measures to improve the intensive wheat-maize system in China. Ambio. 2010;**39**(5-6):376-384. DOI: 10.1007/s13280-010-0076-6

[18] Zhang L, Zhang X, Gao Q, Yan L. Nitrogen application effect on maize yield, NH3, and N2O emissions in northeast China by meta-analysis. Agronomy. 2023;**13**(6):1479. DOI: 10.3390/agronomy13061479

[19] Chen XX, Jiao JY, Cao WQ, Yu BG, Liu YM, Zou CQ. A sustainable phosphorus management in agriculture: Assessing trade-offs between human health risks and nutritional yield regarding heavy metals in maize grain. Environmental Research. 2022;**203**:111792. DOI: 10.1016/j.envres.2021.111792

[20] Ma W, Ma L, Li J, Wang F, Sisák I, Zhang F. Phosphorus flows and use efficiencies in production and consumption of wheat, rice, and maize in China. Chemosphere. 2011;**84**(6):814-821. DOI: 10.1016/j.chemosphere.2011.04.055

[21] Zingore S, Adolwa IS, Njoroge S, Johnson JM, Saito K, Phillips S, et al. Novel insights into factors associated with yield response and nutrient use efficiency of maize and rice in sub-Saharan Africa. A review. Agronomy for Sustainable Development. 2022;**42**(5):82

[22] Zhang J, Jiang X, Xue Y, Li Z, Yu B, Xu L, et al. Closing yield gaps through soil improvement for maize production in coastal saline soil. Agronomy. 2019;**9**(10):573

[23] Zhiqing J, Daokuo G, Xilian Z, Hua C. Assessing the potential impacts of global climate change on maize production in China. Zuo wu xue bao. 1 Jan 1996;**22**(5):513-524

[24] Jiang W, Zhang W, Ding Y, et al. Effects of narrow plant spacing on root distribution and physiological nitrogen use efficiency in summer maize. The Crop Journal. 2013;**1**(1):77-83. DOI: 10.1016/j.cj.2013.07.011

[25] Meng QF, Yue SC, Hou P, Cui ZL, Chen XP. Improving yield and nitrogen use efficiency simultaneously for maize and wheat in China: A review. Pedosphere. 2016;**26**(2):137-147

[26] Sabri N, Kassim NS, Ibrahim S, Roslan R, Abu Mangshor NN, Ibrahim Z. Nutrient deficiency detection in maize (Zea mays L.) leaves using image processing. IAES International Journal of Artificial Intelligence (IJ-AI). 2020;**9**(2):304-309. DOI: 10.11591/ijai.v9.i2.pp304-309

[27] Jat RK, Bijarniya D, Kakraliya SK, Sapkota TB, Kakraliya M, Jat ML. Precision nutrient rates and placement in conservation maize-wheat system: Effects on crop productivity, profitability, nutrient-use efficiency, and environmental footprints. Agronomy. 2021;**11**(11):2320. DOI: 10.3390/agronomy11112320

[28] Nunes da Silva M, Machado J, Osorio J, Duarte R, Santos CS. Non-essential elements and their role in sustainable agriculture. Agronomy. 2022;**12**(4):888

[29] Song XD, Liu F, Wu HY, Cao Q, Zhong C, Yang JL, et al. Effects of long-term K fertilization on soil available potassium in east China. Catena. 2020;**188**:104412

[30] Nisa ZU. Comparison of different potassium application methods for maize (Zea mays L.) growth under salt-affected soils. Journal of Pure and Applied Algebra (JPAA). 2021;**6**(1):44-53

[31] Augustine R, Kalyansundaram D. Agronomic biofortification through micronutrient management in maize: A review. Journal of Applied and Natural Science. 2020;**12**(3):430-437

[32] Anas M, Liao F, Verma KK, Sarwar MA, Mahmood A, Chen ZL, et al. Fate of nitrogen in agriculture and environment: Agronomic, eco-physiological and molecular approaches to improve nitrogen use efficiency. Biological Research. 2020;**53**(1):1-20

[33] Sarwar N, Farooq O, Wasaya A, Hussain M, El-Shehawi AM, Ahmad S, et al. Integrated nitrogen management improves productivity and economic returns of wheat-maize cropping system. Journal of King Saud University-Science. 2021;**33**(5):101475

[34] Zone PP. Effect of Soil Test Values and Fertilization on Corn, Soybean and Wheat Tissue Phosphorus and Potassium Concentrations [Master's thesis]. Ohio, USA: Ohio State University; 2019 [cited 2023 Aug 12]. Available from: http://rave.ohiolink.edu/etdc/view?acc_num=osu1546448067902974

[35] Preston CL, Ruiz Diaz DA, Mengel DB. Corn response to long-term phosphorus fertilizer application rate and placement with strip-tillage. Agronomy Journal. 2019;**111**(2):841-850

[36] Jatav HS, Sharma LD, Sadhukhan R, Singh SK, Singh S, Rajput VD, et al. An overview of micronutrients: Prospects and implication in crop production. In: Sadhukhan R, editor. Plant Micronutrients: Deficiency and Toxicity Management. 1st ed. Chapter 1. New York City: Springer International Publishing; 2020. pp. 1-30

[37] Kuyah S, Sileshi GW, Nkurunziza L, Chirinda N, Ndayisaba PC, Dimobe K, et al. Innovative agronomic practices for sustainable intensification in sub-Saharan Africa. A review. Agronomy for Sustainable Development. 2021;**41**:1-21

[38] Gram G, Roobroeck D, Pypers P, Six J, Merckx R, Vanlauwe B. Combining organic and mineral fertilizers as a climate-smart integrated soil fertility management practice in sub-Saharan Africa: A meta-analysis. PLoS One. 2020;**15**(9):e0239552

[39] Sharma V, Singh MJ, Khokhar AK. Productivity, nutrient uptake and soil properties as influenced by integrated nutrient management in maize-wheat cropping system under rainfed conditions of sub-montane Punjab. Agricultural Research Journal. 2020;**57**:839-847

[40] Fageria NK, Baligar VC, Li YC. The role of nutrient efficient plants in improving crop yields in the twenty first century. Journal of Plant Nutrition. 2008;**31**(6):1121-1157

[41] Wu P, Liu F, Li H, Cai T, Zhang P, Jia Z. Suitable fertilizer application depth can increase nitrogen use efficiency and maize yield by reducing gaseous nitrogen losses. Science of The Total Environment. 2021;**781**:146787

[42] Gao C, El-Sawah AM, Ali DFI, Alhaj Hamoud Y, Shaghaleh H, Sheteiwy MS. The integration of bio and organic fertilizers improve plant growth, grain yield, quality and metabolism of hybrid maize (Zea mays L.). Agronomy. 2020;**10**(3):319

[43] van Bruggen AH, Goss EM, Havelaar A, van Diepeningen AD, Finckh MR, Morris JG Jr. One Health-Cycling of diverse microbial communities as a connecting force for soil, plant, animal, human and ecosystem health. Science of the Total Environment. 2019;**664**:927-937

[44] Shi W, Zhang Q, Xie R, Sheng K, Tan J, Wang Y. Blending loss-control and normal urea maximizes nitrogen utilization of summer maize by mitigating ammonia volatilization and nitrate leaching. Plant and Soil. 2023:1-17

[45] Adnan M. Role of potassium in maize production: A review. Open Access Journal of Biomedical Science Research. 2020;**3**(5):1-4

[46] Robert PC. Precision agriculture: A challenge for crop nutrition management. In: Progress in Plant Nutrition: Plenary Lectures of the XIV International Plant Nutrition Colloquium: Food Security and Sustainability of Agro-Ecosystems through Basic and Applied Research. Netherlands: Springer; 2002. pp. 143-149

[47] Cassman KG, Dobermann A, Walters DT. Agroecosystems, nitrogen-use efficiency, and nitrogen management. AMBIO: A Journal of the Human Environment. 2002;**31**(2):132-140

[48] Tsouros DC, Bibi S, Sarigiannidis PG. A review on UAV-based applications for precision agriculture. Information. 2019;**10**(11):349

[49] Gorai T, Yadav PK, Choudhary GL, Kumar A. Site-specific crop nutrient management for precision agriculture–A review. Current Journal of Applied Science and Technology. 2021;**40**(10):37-52

[50] Argento F, Anken T, Abt F, Vogelsanger E, Walter A, Liebisch F. Site-specific nitrogen management in winter wheat supported by low-altitude remote sensing and soil data. Precision Agriculture. 2021;**22**:364-386

[51] Meena BP, Biswas AK, Singh M, Chaudhary RS, Singh AB, Das H, et al. Long-term sustaining crop productivity and soil health in maize–chickpea system through integrated nutrient management practices in Vertisols of central India. Field Crops Research. 2019;**232**:62-76

[52] Das D, Dwivedi BS, Meena MC, Singh VK, Tiwari KN. Integrated nutrient management for improving soil health and crop productivity. Indian Journal of Fertilisers. 2015;**11**(4):64-83

[53] Jin H, Zhiqiang Z, Hongwen L, Qingjie W. Development of small/medium size no-till and minimum-till seeders in Asia: A review. International Journal of Agricultural and Biological Engineering. 2014;7(4):1-2

[54] Giller KE, Witter E, Corbeels M, Tittonell P. Conservation agriculture and smallholder farming in Africa: The heretics' view. Field Crops Research. 2009;**114**(1):23-34

[55] Pittelkow CM, Liang X, Linquist BA, Van Groenigen KJ, Lee J, Lundy ME, et al. Productivity limits and potentials of the principles of conservation agriculture. Nature. 2015;**517**(7534):365-368

[56] Altieri MA, Nicholls CI, Henao A, Lana MA. Agroecology and the design of climate change-resilient farming systems. Agronomy for Sustainable Development. 2015;**35**(3):869-890

[57] Garnett T, Godfray C. Sustainable Intensification in Agriculture. Navigating a Course Through Competing Food System Priorities. Food Climate Research Network and the Oxford Martin Programme on the Future of Food, University of Oxford, UK; 2012. p. 51

[58] Riches CR, Liebman M, Mohler CL, Staver CP. Ecological management of agricultural weeds. Annals of Botany. 2003;**91**(4):499-500. DOI: 10.1093/aob/mcg042

[59] Cornelissen G, Martinsen V, Shitumbanuma V, Alling V, Breedveld GD, Rutherford DW, et al. Biochar effect on maize yield and soil characteristics in five conservation farming sites in Zambia. Agronomy. 2013;**3**(2):256-274

[60] Greenwood KL, Mundy GN, Kelly KB. On-farm measurement of the water use and productivity of maize. Australian Journal of Experimental Agriculture. 2008;**48**(3):274-284

[61] Roberts TL. Improving nutrient use efficiency. Turkish Journal of Agriculture and Forestry. 2008;**32**(3):177-182

[62] Boesch DF, Brinsfield RB, Magnien RE. Chesapeake Bay eutrophication: Scientific understanding, ecosystem restoration, and challenges for agriculture. Journal of Environmental Quality. 2001;**30**(2):303-320

[63] Li C, Li Y, Li Y, Fu G. Cultivation techniques and nutrient management strategies to improve productivity of rain-fed maize in semi-arid regions. Agricultural Water Management. 2018;**210**:149-157

[64] Hayati D, Ranjbar Z, Karami E. Handbook of measuring agricultural sustainability. Biodiversity, Biofuels, Agroforestry and Conservation Agriculture. 2011:73-100

[65] Westerman PW, Bicudo JR. Management considerations for organic waste use in agriculture. Bioresource Technology. 2005;**96**(2):215-221

[66] Oliveira L, Costa M, Sousa H, Blum J, Silva G, Abreu M, et al. Characterization of organic wastes and effects of their application on the soil. The Journal of Agricultural Science. 2018;**10**(6):291. DOI: 10.5539/jas.v10n6p291

Chapter 4

Sustainable Maize Production and Carbon Footprint in Arid Land Context: Challenges and Perspectives

El Khalfi Chaima, Harkani Assia, Ouhemi Hanane, Benabdelouahab Tarik and Elaissaoui Abdellah

Abstract

Maize is a versatile crop that serves as a staple food for millions of people and provides various raw materials. Its adaptability to different climates and potential makes it economically valuable. However, the ongoing emissions of greenhouse gases pose significant challenges to sustain maize production. Sustainable agricultural practices are crucial to mitigate greenhouse gas emissions and reduce carbon footprints. Conservation tillage practices based on no-till promote carbon sequestration, and reduce carbon footprints compared to conventional tillage. These practices potentially improve soil health and water productivity. This chapter explores various aspects to sustain maize production, with a focus on conventional and conservation tillage systems, engineering technologies, carbon footprint reduction. It discusses also the challenges and perspectives in achieving sustainable maize production. It begins with an overview of conventional maize farming, highlighting its practices and challenges. The second section explores the advantages of conservation tillage in maize production. The third part focuses on engineering technologies and precision agriculture tools, as well as remote sensing. In the fourth section, strategies for reducing carbon emissions and adopting clean energy in maize farming are considered. The final part addresses the challenges and perspectives for sustaining maize production, discussing barriers, opportunities, and potential solutions.

Keywords: maize, tillage, no-tillage, water, carbon, footprint

1. Introduction

Maize (*Zea mays* L.) holds immense importance as a major cereal crop, serving as a staple food for over 900 million people in developing countries. Maize earns its esteemed title as the "Queen of Cereals" due to its remarkable demand and impressive adaptability. Maize holds the distinction of being the most abundantly produced cereal worldwide, with a staggering production of 1148 million metric tons. Not only does maize boast the highest average productivity of 5.9 tons per hectare, but its growth rate is also among the most rapid in comparison to other crops [1]. It holds the distinction of being the most significant cereal crop globally in terms of both acreage and production [2]. Additionally, it serves as a valuable raw material for the production of food sweeteners, protein, oil, starch, and even as a fuel source. This versatility is supported by its ability to thrive and adapt to a wide range of climatic conditions worldwide [3]. It is not only a vital food crop but also a significant source of income for many farmers, particularly in developing countries.

Its unique characteristic of being able to be cultivated twice in a year further enhances its economic value. In regions with favorable climatic conditions and appropriate agricultural practices, farmers can harness the potential of double cropping, allowing them to maximize their yields and income from maize cultivation [4].

According to the most recent assessment report by the Intergovernmental Panel on Climate Change, the ongoing emission of greenhouse gases is projected to result in continuous warming and persistent alterations in all aspects of the climate system. This, in turn, increases the probability of experiencing severe, widespread, and irreversible impacts on both human societies and ecosystems. The report highlights the urgent need to address greenhouse gas emissions to mitigate the potential consequences of climate change. Thus, in light of the projected severe and irreversible impacts of climate change, there is an urgent and critical need for sustainable food production systems. Agricultural practices, in conjunction with the combustion of fossil fuels within domestic settings, exert a significant influence on the global carbon (C) and nitrogen (N) cycles. This combined impact has been identified as a potential contributor to the observed global temperature rise [5].

There is a strong recommendation for crop producers to implement efficient management practices in order to reduce GHG emissions and minimize the carbon footprints associated with agricultural products at the farm level [6, 7]. Research has consistently shown that implementing improved agronomic practices can contribute significantly to reducing GHG emissions associated with crop production. These practices not only enhance crop yield but also result in higher inputs of carbon-rich residues, which can contribute to increased carbon storage in the soil [6]. Examples of these effective practices include the use of high-yielding crop varieties, timely management of crop diseases, crop rotation with species that allocate more carbon below ground, and careful nutrient management [7].

Conservation tillage, which encompasses practices such as no-till or reduced tillage along with residue retention, has been widely implemented to enhance soil quality and promote sustainable agriculture. The adoption of no-till (NT) and subsoiling (ST) practices has been proposed as a means to decrease soil organic carbon (SOC) mineralization and promote the accumulation of SOC. These practices involve minimal soil disturbance, leading to enhanced SOC sequestration and reduced carbon dioxide (CO_2) emissions, consequently lowering carbon footprints (CFs). In contrast, conventional tillage methods contribute to higher CO_2 emissions through increased diesel consumption, whereas NT practices result in reduced carbon emissions due to decreased diesel usage. Furthermore, tillage practices influence soil physicochemical properties and have implications for grain and biological yields [8].

There have been very limited studies exploring the C footprint of maize production under variable agronomic practices such as conventional and no-tillage farming systems, especially in the context of Morocco. In this literature review, our objective is to examine maize production within two farming systems: conventional and no-till. We will analyze and compare the carbon footprints associated with these two approaches and aim to draw conclusions regarding the potential carbon footprint reduction achievable through the adoption of no-till farming. By assessing the available literature and research on these farming systems, we will explore the environmental implications of each method and evaluate the carbon footprint gains that can be achieved by transitioning from conventional to no-till maize production. Ultimately, our review aims to provide insights and recommendations on how adopting no-till practices can contribute to sustainable maize production with reduced carbon footprints.

2. Conventional maize production and drawbacks

The overuse of synthetic fertilizers and pesticides leads to soil degradation, nutrient runoff and greenhouse gas emissions. It also disrupts ecosystems, fosters resistance, and poses health and environmental risks. Unsustainable irrigation depletes water resources, causes salinization, and contributes to energy consumption. Solutions include integrated nutrient management, integrated pest management, efficient irrigation techniques, education, policy support, and research.

Tillage plays a crucial role in creating favorable conditions for seedling emergence, development, and root growth by preparing an ideal seedbed. It is considered a critical component of soil management systems. However, it is important to select appropriate tillage practices to ensure optimal crop growth and yield. Inappropriate tillage practices can have detrimental effects on crop performance. Tillage management, as well as the application of chemicals and manure, are important factors that have a significant impact on soil physical properties. Tillage is a practice used to loosen the soil and create a suitable seedbed for plant growth. It plays a crucial role in crop production, contributing up to 20% of the overall factors influencing crop performance [9].

The choice of tillage method has implications for the sustainable utilization of soil resources, as it directly influences soil properties. Deep tillage, in particular, has several benefits. It helps break up compacted soil layers, facilitating improved water infiltration and movement within the soil. This enhanced water penetration allows for better root growth and development, ultimately increasing the potential for crop production [10]. Studies have shown that deep tillage practices, reaching depths of up to 90 cm, have led to increased corn yield [11].

In his study case, Memon et al. [9] compared the effects of different tillage practices, Deep Tillage (DT), Conventional Tillage (CT), and Zero Tillage (ZT), on maize production at the experimental site of National Agriculture Research Center (NARC), Islamabad, Pakistan. The study revealed significant differences among the tillage treatments in terms of seedling emergence, plant height, number of leaves per plant, and grain and dry matter yields. Deep Tillage (DT) exhibited notable advantages over the other treatments. It resulted in a higher seedling emergence percentage, taller plants with more leaves, and the highest grain and dry matter yields. Conventional Tillage (CT) followed suit, demonstrating favorable outcomes in terms of seedling emergence, plant height, leaf number, and yield. Considering the specific soil (loamy soil) and weather conditions (spring season) of the experiment, the findings indicate that Deep Tillage (DT) proved to be the most effective tillage practice for maize production.

Zero Tillage (ZT), although offering potential benefits in certain contexts, was less favorable in this study. These results emphasize the importance of selecting appropriate tillage practices based on specific conditions to optimize maize production outcomes. It is important to note that balancing the benefits and potential drawbacks of tillage management and chemical/manure applications is crucial. While tillage can improve soil conditions and crop productivity, it may also lead to increased soil erosion and loss of organic matter. Likewise, the use of chemicals should be carefully managed to minimize environmental impacts and promote sustainable agricultural practices.

In another study conducted by Orfanou et al. [12] in USA Georgia, they found that Conventional tillage had slightly better yield results but was not statistically different from conservation-tilled plots that lead to the conclusion that conservation tillage could be a good solution for farmers, not only for preserving water but also for achieving acceptable yield results.

Conventional tillage in maize production, while commonly practiced, comes with several drawbacks. It demands significant time, fuel, labor, and water resources, leading to higher production costs. These increased costs can ultimately reduce profits for farmers. Additionally, conventional tillage methods may contribute to higher greenhouse gas emissions compared to alternative options [13]. The constraints and nonsustainability issues mentioned earlier have specific implications. Maize farming often involves intensive tillage practices, which can lead to soil erosion, reduced soil organic matter content, and soil compaction, ultimately impacting the long-term productivity of the land. Additionally, the removal of crop residues and the practice of monocropping in maize cultivation further contribute to the nonsustainable aspects of conventional agricultural systems [14]. To ensure sustainable maize production, it is crucial to address these challenges and adopt alternative farming practices such as conservation tillage.

In a study conducted in South Korea under intensive conventional cultivation, [15] concluded that the carbon footprint (CF) of maize production is largely influenced by nitrogen (N) in chemical fertilizers and the use of organic fertilizers. Both types of fertilizers significantly contribute to CF and carbon efficiency. Sustainable practices that prioritize high yields and low GHG emissions are associated with greater sustainability. South Korea's maize production demonstrates relatively low CF and GHG emissions on a global scale. The study highlights the positive correlation between nitrogen use, chemical and organic fertilizers, and the carbon footprint of agriculture (CFA) and carbon footprint intensity (CFI). This emphasizes the importance of proper management and selecting suitable land management systems, especially in the context of climate change. By implementing effective strategies and informed decision-making, it is possible to reduce GHG emissions and promote sustainable development in maize farming.

3. Maize-based conservation tillage and benefits

Conservation tillage plays a significant role in sustainable maize production by promoting soil health, reducing erosion, improving water retention, and minimizing the environmental impact of farming practices.

Maize cultivation can be achieved without the need for primary tillage through a practice known as no-till farming. This approach offers several advantages, including reduced cultivation costs and improved efficiency in resource utilization. To ensure successful crop establishment, it is essential to maintain optimal soil moisture during sowing. Additionally, the proper placement of seeds and fertilizers in bands using a zero-till seed-cum-fertilizer planter with a suitable furrow opener, taking into account the soil texture and field conditions, is crucial [2].

A study conducted in Zimbabwe in a semi-arid climate examined the impact of conservation tillage on maize production. The objective of the study was to assess the maize yield advantage associated with conservation tillage compared to conventional tillage, which represented the farmers' practice in the region. The researchers aimed to provide insights into the potential benefits of introducing conservation tillage as a technology for maize production in the semi-arid conditions of Zimbabwe and compare the efficacy of conventional tillage and conservation tillage methods in terms of maize yield. When comparing the performance of various tillage methods, it is important to acknowledge that for any alternative system to be considered viable, its yield should be equal to or higher than that of conventional tillage in the short term. Additionally, it is crucial to consider the resource constraints faced by smallholder farmers during the adoption of alternative tillage practices. They evaluated eight tillage experiments conducted between 1984 and 2008. Nyakudya et al. [16] found results that showed Conservation tillage methods demonstrated slight but noteworthy yield advantages in regions with less than 500 mm of rainfall. In cases where grain yields reached 2.5 tons per hectare and the rainfall was below 500 mm, the adoption of 1.0 m tied ridging resulted in an additional 144 kg per hectare, while mulch ripping contributed an extra 344 kg per hectare compared to conventional tillage practices. These findings highlight the potential of conservation tillage methods to enhance maize yields in areas with limited rainfall, albeit with modest improvements.

In another research in Western Colorado Research Center in USA, Keshavarz et Dekamin 2022 evaluated the sustainability of maize production by comparing four different tillage systems: conventional tillage with moldboard plow (MP), conventional tillage with chisel plow (CP), strip-tillage (ST), and no-till (NT). The assessment was conducted using life cycle assessment (LCA) and Material Flow Cost Accounting (MFCA) methods. By considering the entire production process, including energy and material wastage, a more comprehensive understanding of the hidden costs of production was obtained. The results showed that the total annual energy input varied among the tillage systems, with NT having the lowest energy demand. NT also exhibited improved energy efficiency and yield increase. The economic analysis revealed that eliminating negative products in corn production could significantly increase farmers' net benefit. Environmental impact assessments indicated that NT and MP performed better than CP and ST in most categories, with NT showing the best performance in terms of global warming potential, acidification, and eutrophication. Overall, NT proved to be the most sustainable option for corn production, followed by the MP system, considering energy, economic, and environmental indicators.

4. Practices and engineering technologies for sustainable maize production

Sustainable maize production is of utmost importance in ensuring food security, environmental preservation, and the well-being of farming communities. To achieve sustainability, it is crucial to adopt the best practices and technologies that optimize resource utilization, reduce environmental impact, and enhance the resilience of maize farming systems. During the last decades, several farming innovations have been tested on cereal crops to improve water and energy use efficiencies and increase yields of biomass and grains. Dokyi et al. [17] stated that the adoption of Improved Seed and Management Technologies (ISMT) has a significant positive impact on technical efficiency. The ISMT adoption resulted in a notable increase of the efficiency to show an actual improvement of 16%. Consequently, the maize productivity showed a substantial boost, rising by 33.8% as a result of ISMT adoption. This study recommends the widespread dissemination of improved maize seeds to farmers. The transformation of corn farming over the past two decades has been fueled by the rapid adoption of new technologies and advancements in seed breeding. A comprehensive analysis (ARMS survey conducted from 1996 to 2016) reveals the significant impact of these innovations on yield changes in intensive corn production.

Otherwise, the advancements in genetically engineered seeds allowed farmers to optimize their practices and achieve higher yields. With the ability to plant corn seeds more densely and at an earlier stage in the growing season, farmers maximized the crop's growth potential. Additionally, the improved pest resistance and drought tolerance provided by genetically engineered seeds opened up profitable production opportunities in different pedo-climatic contexts. These changes were not limited to planting practices alone, as the increased adoption of drought-tolerant and insect-resistant seeds prompted adjustments in irrigation and chemical applications. Over the course of two decades, the percentage of corn acres planted with single-pest-resistant varieties containing proteins from *Bacillus thuringiensis* (Bt) increased from 2% in 1996 to 78% by 2016. Similarly, herbicide-tolerant varieties, enabling efficient weed management, saw a remarkable area increase from 3% in 1996 to 84% in 2016 [18]. However, adoption of genetically engineered seed varieties improved substantially productivity of conventional farming but the sustainability of this production system cannot maintained as different problems of soil health, soil physic, pest resistance; herbicide tolerance and chemical pollution kept unsolved in a sustainable way.

4.1 Practices for better maize crop establishment under no-till

Several practices were proved to improve maize crop establishment for more sustainability under no-till cropping system. For a successful introduction of no-till farming method, farmers cannot sense an initial benefit without starting by fixing problem of soil compaction as a common issue of intensive agriculture. Compaction can be attributed to various field operations related to soil-machine interactions due to use of heavy machinery and equipment and to animal trampling. These activities can result in damage to the soil structure, which is crucial for the soil's ability to retain and drain water, nutrients, and air necessary for plant root functions. Compacted soil restricts root growth and can lead to reduced water infiltration, poor nutrient availability, and inadequate oxygen levels for plant roots. Several researches showed that compaction constitutes a systematic problem of irrigated cropping systems due to difficulty of traffic management with reference to soil practicability and soil plasticity. Olubanjo et al. [12] conducted a study in Nigeria to show the response of maize crop to compacted soil under relatively stable environmental conditions. They find that high soil strength resulting from compaction lead to reduced yield production. However, the negative impact of compaction on yield seems to be mitigated when there is an abundant water supply, although certain treatments with lower soil strengths experienced further reduction in yield due to water stress. Additionally, increased soil compaction was found to have a negative influence on plant nutrient uptake. According to this study, maize plants should not be cultivated in soils with a penetration resistance more than 2.0 MPa.

Methods of chiseling and tillage of deep soil layers are of great importance to break hardpans and to alleviate soil compaction prior to cultivation of maize under conventional tillage. Such methods are also primordial for a successful start of producing maize under conservation tillage. In fact, it is essential to address soil compaction through proper management practices such as chiseling before no-tillage and use of adapted no-till seeders for a better maize crop establishment. The conservative best practices help farmers to guarantee a sustainable maize production when the maize crop establishment is good to show consistent biomass and grains yields during the start years of the conservative practices (**Figure 1**).

To enhance maize crop productivity and improve farmers' profitability, there is a significant focus on implementing alternative methods and technologies to promote conservative practices. These efforts aim to mitigate the negative impact of traditional cropping systems and have resulted in the development of various resource conservation technologies. Considering the importance of conserving natural resources, it is crucial to prioritize the widespread adoption of cost-effective and environmentally friendly crop management practices. These include techniques such as ridge and furrow, conventional flatbed, and raised-bed planting [20].

The ridge and furrow planting method involves creation of raised ridges and sunken furrows for a better crop establishment. This method offers several benefits for crop growth. The ridges provide better drainage and aeration for the plants, reducing the risk of waterlogging. The furrows help to channel water and prevent excessive runoff, improving water distribution and conservation.

The conventional flatbed planting method can be prepared by leveling the soil surface to create a flat and even bed for planting. In this method, the entire field is tilled and smoothed to achieve a uniform surface for easier planting, cultivation, and harvesting operations.

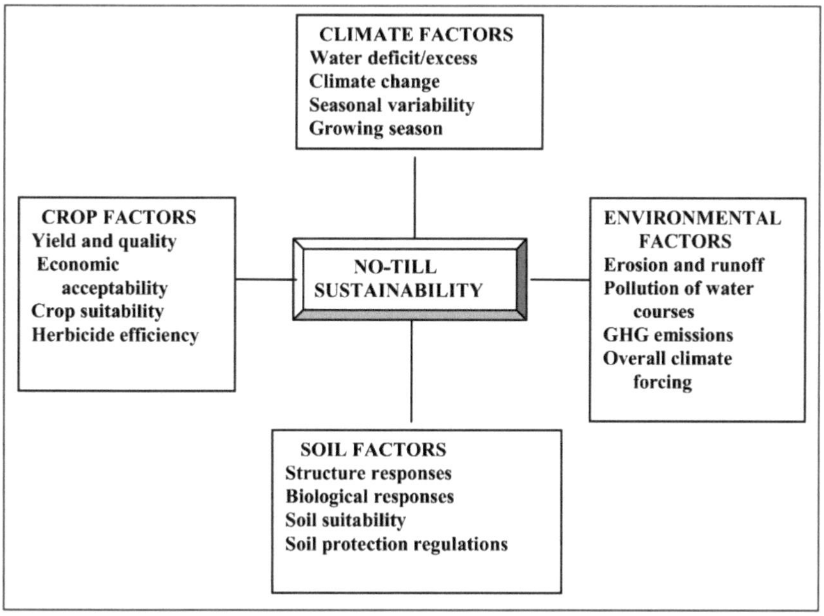

Figure 1.
Factors affecting no-till production system sustainability [19].

The zero tillage raised-bed planting method involves creation of elevated planting beds above the ground level. The raised beds are typically formed by mounding soil or using specialized equipment to shape them.

Saad et al. [21] conducted a study in India to find that energy use in tillage is influenced by different tillage and crop establishment methods, as well as residue management practices. The zero tillage with raised-bed establishment (ZTB) consumed approximately 8% less energy compared to conventional tillage based on flatbed planting (CTF) in a maize-wheat cropping system. This reduction in energy consumption in ZTB was due to energy savings in land preparation, sowing, and irrigation activities.

Pooja et al. [20] also examined the impact of different planting methods on weed population, yield improvement, water management, and weed control in maize production. The results indicate that raised beds have lower weed populations and offer advantages such as better water management and higher yields compared to flat beds. Stale seedbed practices also prove effective in reducing weed density. Bed planting methods, particularly raised beds, demonstrate higher soil microbial biomass carbon and have a significant positive effect on crop growth and yields. Studies conducted by Jat et al. and Singh et al. show that raised-bed systems outperform conventional and zero tillage systems in terms of maize yield. Overall, the research suggests that raised-bed planting is the most effective method for minimizing weed population and enhancing crop performance [20].

4.2 Digital monitoring of crop performance for sustainable maize production

There are several technologies that contribute to sustainable maize production. For example, Soil–Plant Analysis Development (SPAD) meter technology. It has emerged as a valuable tool in the field of agriculture. This technology has gained significant attention, particularly in the context of optimizing nitrogen fertilizer applications in corn (*Zea mays* L.) production. The SPAD meter is a handheld device that measures the chlorophyll content of plant leaves, providing an indication of their nitrogen status. The use of SPAD meter technology offers several advantages for corn producers. By providing a quick and nondestructive assessment of leaf chlorophyll levels, it enables farmers to monitor the nitrogen status of their crops in real-time. This information is crucial for making informed decisions about nitrogen fertilizer applications, ensuring that the crops receive adequate nutrients for optimal growth and yield.

Farmers often opt for high nitrogen (N) rates to maximize corn yield, highlighting the need to determine optimal N quantities for promoting efficient farming practices that increase yield and crop profitability while minimizing resource wastage. Striking the right balance is crucial, as excessive N application poses a challenge for both farmers and environmentalists in safeguarding groundwater against nitrate contamination. By adopting appropriate N management strategies, farmers can mitigate the potential negative impacts of excessive N use, reduce environmental risks, and contribute to sustainable maize production.

Rhezali et al. [22] conducted a study in 2014 and 2015 to show that it is possible to explore the relationship between absolute SPAD values and leaf nitrogen concentration, focusing on early corn growth stages such as V6, V8, V10, and V12. Three experiments were conducted to examine the effects of six nitrogen (N) treatments applied at early growth stages of corn. The results indicated a significant linear relationship between corn leaf N concentrations and absolute SPAD values, with an

R^2 value of 0.80 ($p < 0.05$). Interestingly, the average optimal corn leaf N concentration decreased as the corn progressed through its growth stages.

Ensuring accessibility of the absolute SPAD method is crucial for its practical application by farmers. The absolute SPAD method, which has shown a significant linear relationship between corn leaf nitrogen concentrations and SPAD values, holds promise for aiding farmers in making informed decisions about nitrogen applications.

Otherwise, satellite imagery and remote sensing techniques have revolutionized the monitoring of maize crops, providing indispensable tools for farmers. Through the development of innovative algorithms and models, researchers have harnessed satellite data to extract valuable information for crop management. These insights include crop yield prediction, disease detection, and analysis of nutrient deficiencies [15, 23]. By leveraging satellite-based monitoring systems, farmers can make data-driven decisions to enhance their crop management practices. Furthermore, satellite and drone technologies have also facilitated the implementation of variable rate application of inputs in maize production. By mapping field variability, these technologies optimize the application of fertilizers, herbicides, and pesticides, resulting in reduced costs and environmental impacts while maximizing yields. The implementation of variable rate application ensures efficient resource utilization and promotes sustainable maize production [24, 25].

Satellite and drone have also been used for crop imaging to provide farmers with detailed information on the health and vigor of their maize crops. By employing multispectral and thermal sensors, farmers can assess crop stress, monitor water use efficiency, detect nutrient deficiencies, and quantify vegetation indices such as NDVI (Normalized Difference Vegetation Index). These assessments enable farmers to take proactive measures to mitigate potential risks and optimize maize production [26, 27].

In addition to satellites, drones equipped with sensors and cameras have emerged as valuable tools for precise data collection in maize fields. Drones capture high-resolution images that enable the measurement of plant height, identification of nutrient deficiencies, and detection of pests and diseases. These images also contribute to the creation of yield maps, providing farmers with detailed information for optimizing fertilization, irrigation, and pest control strategies [28, 29]. The integration of these data-driven insights empowers farmers to make informed decisions, resulting in improved maize production.

The combination of satellite and drone data with crop modeling and decision support systems has further enhanced the accuracy of maize growth prediction and management. Researchers are actively developing models that incorporate climatic data, satellite imagery, soil characteristics, and management practices. These integrated systems optimize irrigation scheduling, planting dates, and fertilizer application, ultimately enabling farmers to achieve better yields [30–32]. By leveraging these tools, farmers can confidently make decisions based on accurate predictions and optimize their maize production.

Sharifi [33] implemented a model using Near-Infrared Reflectance (NIR) and Red-edge bands in vegetation indices to precisely predict maize nitrogen uptake in three different sites and various conditions. He stated that maize growers can have a good opportunity to map nitrogen uptake for improving nitrogen use efficiency in their field. Use of spectral information of Sentinel-2 satellite data for estimating maize nitrogen uptake served as an efficient tool to optimize fertilizer management in irrigation-based intensive cropping systems.

4.3 Contribution of precision irrigation technologies for sustainable maize production

Smart Irrigation and Internet of Things (IoT) technologies consistently contributed to improve water and maize crop productivity. The power of the Internet of Things (IoT) can be used with sensors to monitor various factors like soil moisture levels, weather conditions, and plant water requirements. By collecting real-time data, smart irrigation systems can optimize water use during the cropping system to ensure precise irrigation schedules for more water use efficiency and productivity [34, 35]. Several approaches integrating use of IoT and sensor network have been implemented to efficiently collect and analyze data for promoting more sustainability in the irrigated cropping systems. Use of processed data at the edge server and transferred to the main IoT server is a real-time process of great utility to continuously manage the crops water requirements using only an Android smartphone application [36–38]. By implementing precise irrigation based on soil moisture sensors and IoT, maize producers can achieve higher yields while optimizing the use of resources such as fertilizers, water, and seeds [39, 40]. The comparison between precise irrigation using sensors and traditional flood irrigation showed that it is possible to conserve water by 50% and increase crop yield by 35% [41]. Integration of IoT technology is also of great importance to adapt for monitoring irrigation data for diverse crops. Singh et al. [26] evaluated an automated irrigation system for Maize, Paddy and Wheat crops to monitor soil moisture and soil temperature and transmit data to a cloud system for digital control of pump to efficiently satisfy the irrigation requirements. By considering Maize as the most important cereal crop worldwide [42], emerging sensors technologies can be of great importance to implement powerful tools helping for more sustainability in producing maize silage and grains. It helps farmers to implement decisional tools based on real-time data [43]. Sharifnasab et al. [44] tested smart irrigation for producing maize grain to show possibility of using only 40% of the farm's moisture discharge capacity. Compared to conventional practice of using meteorological data to guide irrigation decisions, the implementation of a smart irrigation system resulted in accelerated plant growth, earlier harvesting, and reduced water use (from 8839.5 to 5675.67 m^3/ha) for more grain yield and water productivity [44]. Kumar et al. [45] evaluated an irrigation method based on IoT to monitor soil moisture monitoring with reference to use of evapotranspiration-based strategy to manage sweet corn irrigation. The first IoT-based method implemented for two irrigation regimes of 43.5% and 34.8% of the soil field capacity (FC) is compared to the evapotranspiration method (ETc 100%) with 80% of FC. They find that the IoT method based on regime of 43.5% resulted in an increase yield of 12% and water savings of 11% compared to the ETc 100% irrigation method. Asiimwe et al. [46] compared and evaluated sweet corn yield, biomass, water productivity, and other morphometric characteristics based on irrigation scheduling using the irrigation amounts estimated from ET (60%, 90%, and 120% of ETc) and SM irrigation regimes (25%, 30%, and 35% of soil moisture) on sweet corn. The results showed that the average soil moisture levels using both treatments soil moisture (SM35%) and evapotranspiration (ET120%) were identic to show that irrigation can be reduced by 8% for the same grain yield and the highest irrigation level can result in an increase of fresh cob weight by 27%. Such smart irrigation innovations can help to elevate productivity levels while also ensuring sustainable agricultural practices [47]. Considered as a key component of precision farming, this advanced technology

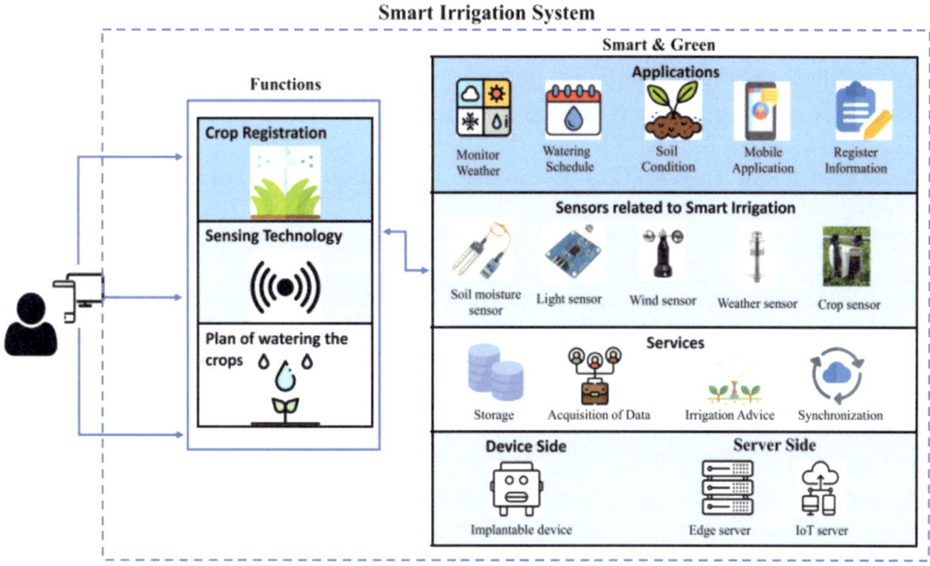

Figure 2.
Smart irrigation system structures (From www.flaticon.com).

is becoming affordable to be adopted by small-irrigated farms to optimize water productivity and enhance crop yield through implementation of irrigation best practices (**Figure 2**) [48].

4.4 Crop growth modeling for sustainable maize production

Maize crop as other crops is subjected to effect of the current meteorological conditions of climate change. Which affect negatively the yield of the crop. For this, growth simulation models are used to simulate different scenarios under the actual climate change [49, 50]. The most affected regions by climate change could be China, Africa, European Union and India, with a maximum decrease in maize yield of 86%, 201%, 71% and 45%, respectively [49]. The major factor affecting the rise in maize yield under climate change is the temperature [50]. The use of models to simulate and forecast the response of maize crop to different environmental conditions are used in several regions as an alternative tool to analyze the response to climate change conditions [49]. However, the simulation models are observed to give mixed results depending on the region and the crop. The parametrization, calibration and validation were found to be the source of uncertainties in model predictions [50]. Different situations could be simulated: those related to optimal conditions with restricted effect of climate conditions (T°, radiation and CO_2), those related to resource availability (water and nutrient), and finally, those related to the reel conditions including all environmental, biological and management variables [49].

Actually, the complexity of the biophysical agricultural system is mathematically formulated by models helped to understand them [49]. The models used to simulate maize production are different in terms of information required, and the end user interface [49]. Climate, plant, soil and crop management are the input data needed by mechanistic models, such as AquaCrop, APSIM, DSSAT-CERES, CropSyst and EPIC [49–51]. Most of the studies revealed that corn yield decreases under climate change projections, due to temperature increase which reduces vegetative period and dry matter production in some regions, while, there are other regions where the conditions of corn crop growth will be favorable (temperate regions) [49]. The use of experiment data is needed to calibrate and validate each model [51]. The calibration of DSSAT-CERES is made for each genotype of maize and estimate the genetic coefficient. In WOFOST model, the calibration is carried out in the different phenological stages [49].

Otherwise, AquaCrop is used as water-driven crop model under varying irrigation and nitrogen level in [51–54]. Model efficiency (E), coefficient of determination (R2), Root Mean Square error (RMSE) and Mean Absolute Error (MAE) Nash–Sutcliffe Efficiency (NSE) were used to test the model performance [51–54]. Appropriate levels of irrigation for maize crop were investigated by using AquaCrop model [51, 53, 55]. The prediction error of the model varied from 2.35 to 27.5% for different levels of irrigation and nitrogen [51]. Some extreme conditions may limit the performance of the model mainly, water stress, excess water and high evaporative demand conditions [52], and the accuracy of the model need more evaluation under field conditions of maize crop. AquaCrop model give good accuracy for field-measured trait for instance soil water, canopy cover, grain yield and total biomass [52, 53, 55]. The methods of field assessment to assessing maize crop yield are expensive, laborious and inaccurate. To overcome this, considerable efforts were made in the development and application of maize crop yield models for yield estimation. Such as the development of the use of models with remote sensing tools [56].

5. Potential of reducing carbon and water footprints for sustainable maize production

5.1 Maize production and carbon footprint

Maize cultivation has traditionally relied on conventional tillage methods involving plowing. However, due to factors such as cost, natural conditions, and environmental concerns, there is an increasing adoption of noninversion systems in modern maize production. These noninversion systems typically involve reduced tillage, where no-till seeders are used for substituting plowing and contributing to sequester more soil carbon. By promoting adoption of the no-tillage system, the seeders put seeds directly into uncultivated soil for low disturbance and high reduction of GHG [57].

In order to protect the soil and the environment, the use of noninversion tillage techniques and the retention of a minimum of 30% of plant residues on the field, known as conservation tillage, are of particular importance. These practices help preserve soil structure and reduce erosion while promoting the conservation of organic matter.

Enhancing the management of soil cultivation practices to promote the sequestration of organic carbon in the soil is crucial for mitigating greenhouse gas (GHG) emissions in agriculture. Recent research conducted by Holka et Bienkowski [57] in Wielkopolska in Poland, has highlighted that the adoption of no-tillage methods combined with substantial crop residue retention can effectively reduce GHG emissions in maize production. Irrespective of the specific tillage system utilized, the process of mineral fertilization emerged as the key contributor to GHG emissions. Developing low-emission technologies necessitates careful consideration of the associated risks, particularly related to nitrogen fertilizer usage. To minimize emissions from agricultural fields and simultaneously reduce raw material consumption in fertilizer production, optimizing fertilization practices becomes essential, taking into account natural constraints and soil conditions, as well as the desired crop productivity levels.

By considering the sequestration of organic carbon (C) in the calculation of greenhouse gas (GHG) emissions, the net carbon footprint (CF net) of grain maize production was found to be significantly reduced. Compared to the baseline CF value, the CF net values were lowered by 42.9% in the conventional tillage (CT) system, 72.1% in the reduced tillage (RT) system, and 78.3% in the no-tillage (NT) system. When GHG emissions were analyzed per ton of maize produced, the inclusion of C sequestration showed the most substantial impact in reducing total GHG emissions in the NT and RT systems, with reductions of 78.3% and 72.1%, respectively. Effective management of maize crop residues, such as leaving larger amounts of residues in the field, played a significant role in preventing C losses promoting its sequestration, and reducing the carbon footprint in maize production.

5.2 Maize production and water footprint

Water footprint (WF) is an indicator that plays a vital role in promoting sustainable maize production by addressing both water consumption and pollution. Maize is a major global crop, and understanding its water footprint is crucial for ensuring responsible water management practices. It provides a new approach for assessing water resource utilization in agriculture.

The WF of crop production serves as a comprehensive indicator that encompasses the various types of water consumption, quantities utilized, and environmental impacts throughout the entire crop growth period [58]. It provides a holistic understanding of water consumption and its associated implications during the process of crop cultivation. The WF takes into account not only the direct water usage by the crops but also the indirect water footprint related to the production and use of inputs such as fertilizers and pesticides.

The WF considers both the blue water footprint (water from surface or groundwater sources) and the green water footprint (rainwater stored in the soil). Additionally, it accounts for the gray water footprint, which refers to the volume of water that is required to assimilate polluted water [59]. In a study conducted by Sun et al. [58] in Beijing, they found that WF had decreasing trends because of the reduced green WF due to the change in climate and the rising temperature and water scarcity, while the gray WF increased because of chemical fertilizers and pesticides. They concluded that the gray WF should be controlled to achieve a sustainable maize production. In another study conducted in Italy by Borsato et al. [60], they affirm that soil conservation tillage systems can reduce the gray WF by 10%. The study focuses on soil tillage systems and variable rate application as means to reduce the gray WF. It emphasizes that the interaction between soil tillage systems and soil management plays a significant role in reducing the gray WF. They found that minimum Tillage with Precision Farming shows lower gray WF values, both in terms of water usage per ton and per hectare. Soil tillage systems combined with variable rate application exhibit a higher reduction in gray WF. To reduce water pollution, prioritizing the reduction of insecticides and herbicides, using chemicals with a lower gray WF, and implementing sustainable soil management practices are recommended.

6. Challenges and perspectives

Maize yields depend on a range of interconnected factors: genetics influence potential productivity, climate affects growth conditions, agronomy practices impact crop health, policies shape resource access, and political stability enables long-term planning. These elements interact to create a complex impact on yields. Improved genetics can enhance resilience, while effective agronomy optimizes potential.

Challenges in sustainable maize production encompass a range of interconnected factors that need to be addressed collectively. These challenges arise from various dimensions, including environmental, social, economic, and technological aspects. A holistic approach is necessary to tackle these challenges effectively and achieve sustainable maize production. According to [61], the sustainability level of maize farming systems is influenced by various socioeconomic characteristics of farmers and their observed climate change adaptations. Factors such as farmers contact with extension services, membership in agricultural organizations, access to credit, farm size, and their adoption of climate change adaptation measures such as on-farm diversification and land use changes were identified as significant driving forces shaping the sustainability of maize farming systems.

One of the primary challenges is the limited adoption of sustainable practices by farmers. Barriers such as lack of awareness, limited access to resources and information, and resistance to change hinder the widespread implementation of sustainable techniques. Overcoming these barriers requires a multifaceted approach that involves promoting awareness through farmer training programs, providing technical support and guidance.

In a paper review conducted by Cairns et al. [62], they concluded that enhancing the nutritional density of maize within farmers' fields is a critical goal. Achieving this requires not only increasing yield but also optimizing the nutritional content of the harvested crop. Another challenge involves promoting the wider adoption of new maize varieties and expediting the replacement of older ones. While the use of increased fertilizers holds the potential to elevate maize yields, recent evidence suggests that the low and fluctuating returns on investment can hinder the uptake of this approach. Moreover, the adoption of novel agricultural technologies is marked by uneven patterns, with female farmers exhibiting lower rates of adoption. Ignoring gender-specific barriers to technology adoption undermines the potential impacts of these advancements. To address these issues, it is imperative to implement strategies that encompass an integrative approach, considering the interconnected nature of the challenges.

Climate change poses another significant challenge to maize production. The impacts of climate change, such as increased frequency and intensity of droughts, floods, and heatwaves, affect the productivity and resilience of maize crops. Adapting maize cultivation to changing climatic conditions and developing resilient maize varieties that can withstand extreme weather events are essential strategies. Furthermore, implementing climate-smart practices like conservation agriculture, water management techniques and newer technologies can help mitigate the adverse effects of climate change on maize production.

Therefore, it is imperative to develop strategies for effectively addressing the challenges posed by climate change and mitigating the detrimental impact of water stress on maize production. Several viable approaches exist for adapting to water stress conditions. The initial approach involves harnessing the diverse genetic pool and identifying sources of drought resistance to develop and release new maize cultivars.

The second approach centers on leveraging biotechnology advancements, utilizing molecular markers and gene transfer techniques to enhance water stress tolerance in maize plants. The third approach emphasizes the refinement of agricultural practices through the integration of meteorological data, ensuring the alignment of farming techniques with prevailing climate conditions. Additionally, adopting appropriate fertigation programs becomes crucial to counteract the adverse consequences of water stress on maize crops [63].

Soil health and nutrient management present ongoing challenges for sustainable maize production. Issues such as soil erosion, nutrient depletion, and imbalanced fertilizer use can degrade soil fertility and reduce crop productivity. Implementing soil conservation practices, including cover cropping, crop rotation, and precision nutrient management, can help address these challenges and improve soil health over the long term. Efficient water management is crucial for sustainable maize production, especially in regions facing water scarcity such as Morocco in the last decades. Challenges arise from inefficient irrigation practices, water competition, and limited access to water resources. Adopting precision irrigation techniques, promoting water-saving technologies, and implementing sustainable water management practices can optimize water use efficiency and mitigate the risks associated with water scarcity.

It requires collaborative efforts to address these multifaceted challenges that should involve farmers, researchers, policymakers, and other stakeholders. Enhancing knowledge and capacity building is a key component in promoting sustainable maize production. Providing training programs, farmer field schools, and knowledge-sharing platforms can help farmers, extension services, and stakeholders stay updated on best practices, technological innovations, and sustainable farming techniques.

Policy support from governments and policymakers is crucial for creating an enabling environment for sustainable maize production. This can include providing financial incentives, subsidies for sustainable inputs, and creating market opportunities for sustainably produced maize. Policy interventions can play a significant role in driving the adoption of sustainable practices at a larger scale.

Continued investment in research and innovation is essential to advance sustainable maize production. For example, developing improved maize varieties with traits like drought tolerance, disease resistance, and high nutritional value, researchers can enhance productivity and sustainability. Promoting research on sustainable farming techniques, precision agriculture, and climate-smart practices can unlock new approaches and technologies that contribute to sustainable maize production.

Partnerships and collaboration among various stakeholders are vital for driving sustainable maize production. Collaboration among farmers, researchers, government agencies and private sector actors fosters knowledge exchange, technology transfer, and collective action. Working together, stakeholders can address shared challenges, promote sustainable practices, and achieve the common goal of sustainable maize production.

According to the benefits of implementing smart irrigation and IoT technology, certain challenges have been evoqued for promoting a cost-effective digitalization to improve sustainability of irrigated maize cropping systems:

Inefficient fertilizer practices and the demand for irrigation water contribute to environmental impacts, such as increased greenhouse gas (GHG) emissions and poor water quality, which result in business risks in corn production. Efforts are needed to limit GHGs and manage environmental threats by promoting environmentally friendly technologies, practices, and production products, and by encouraging investments in green technologies. Field screening and monitoring are necessary to quickly identify any issues, such as plant emergence problems, nitrogen shortages, insect infestations, epidemics, weed problems, and the effects of water stress.

Utilization of wireless data collection holds promises in enabling farmers to optimize water usage. However, implementing these components underground presents challenges. One such challenge arises when burying antennas that transmit sensor data in soil, as their performance characteristics undergo significant variations based on the moisture content of the soil.

It is also important to consider that farmers typically operate on narrow profit margins, making IoT systems potentially unaffordable for them. Consequently, in order for these systems to have a viable commercial future, there should be a decreasing trend in the cost of IoT devices and overall system implementation.

Author details

El Khalfi Chaima[1], Harkani Assia[1], Ouhemi Hanane[2], Benabdelouahab Tarik[2] and Elaissaoui Abdellah[1*]

1 Regional Center of Agronomic Research of Settat, National Institute of Agronomic Research, Rabat, Morocco

2 Department of Natural Resource and Environment, National Institute of Agronomic Research, Rabat, Morocco

*Address all correspondence to: abdellah.elaissaoui@inra.ma

References

[1] Bamboriya SD, Jat SL, Shreelatha D, Mahala DM, Rakshit S. Mechanized Maize Production for Enhanced Productivity and Profitability. Ludhiana: Indian Institute of Maize Research; 2020

[2] Jaidka M, Bathla S, Kaur R. Improved Technologies for Higher Maize Production. In Maize-Production and Use. London, UK: IntechOpen; 2019

[3] Gezahegn AM. Role of integrated nutrient management for sustainable maize production. International Journal of Agronomy. 2021;**2021**:9982884. DOI: 10.1155/2021/9982884

[4] Cheema M, Farhad W, Saleem M, Khan H, Munir A, Wahid M, et al. Nitrogen management strategies for sustainable maize production. Crop & Environment. 2010;**1**:49-52

[5] IPCC—Intergovernmental Panel on Climate Change. Climate Change Group I to the Fourth Assessment Report of the Intergovernmental Panel on Climate Change. Cambridge: Cambridge University Press; 2007

[6] Gan Y, Liang C, Hamel C, Cutforth H, Wang H. Strategies for reducing the carbon footprint of field crops for semiarid areas. A review. Agronomy for Sustainable Development. 2011;**31**(4):643-656. DOI: 10.1007/s13593-011-0011-7

[7] Ma BL, Liang BC, Biswas DK, Morrison MJ, McLaughlin NB. The carbon footprint of maize production as affected by nitrogen fertilizer and maize-legume rotations. Nutrient Cycling in Agroecosystems. 2012;**94**(1):15-31. DOI: 10.1007/s10705-012-9522-0

[8] Liu Q-Y, Xu C-T, Han S-W, Li X-X, Kan Z-R, Zhao X, et al. Strategic tillage achieves lower carbon footprints with higher carbon accumulation and grain yield in a wheat-maize cropping system. Science of the Total Environment. 2021;**798**:149220. DOI: 10.1016/j.scitotenv.2021.149220

[9] Memon SQ, Mirjat MS, Mughal AQ, Amjad N. Effect of conventional and non-conventional tillage practices on maize production. Pakistan Journal of Agriculture, Agricultural Engineering, Veterinary Sciences. 2013;**29**(2):155-163

[10] Bennie ATP, Botha FJP. Effect of deep tillage and controlled traffic on root growth, water-use efficiency and yield of irrigated maize and wheat. Soil and Tillage Research. 1986;**7**(1-2):85-95. DOI: 10.1016/0167-1987(86)90010-3

[11] Versa EC, Chang SK, Abolaji JQ, Farquhar DA, Olsen FJ. Effect of deep tillage on soil physical characteristics and corn (*Zea mays* L.) root growth and production. Soil and Tillage Research. 1997;**43**:219-228

[12] Orfanou A, Pavlou D, Porter WM. Maize yield and irrigation applied in conservation and conventional tillage at various plant densities. Water. 2019;**11**(8):1726. DOI: 10.3390/w11081726

[13] Johansen C, Haque ME, Bell RW, Thierfelder C, Esdaile RJ. Conservation agriculture for small holder rainfed farming: Opportunities and constraints of new mechanized seeding systems. Field Crops Research. 2012;**132**:18-32. DOI: 10.1016/j.fcr.2011.11.026

[14] Adhikary S, Biswas B, Priya A. Conservation agriculture: An efficient

tool to overcome the drawbacks of conventional agricultural system towards sustainable crop production. International Journal of Current Microbiology and Applied Sciences. 2020;**9**:1333-1340. DOI: 10.20546/ijcmas.2020.907.154

[15] Li Z, Yang Y, So N, Lee J-I. Carbon footprint of maize planting under intensive subsistence cultivation in South Korea. International Journal of Climate Change Strategies and Management. 2023;**15**(3):301-321. DOI: 10.1108/IJCCSM-12-2021-0141

[16] Nyakudya IW, Stroosnijder L. Conservation tillage of rainfed maize in semi-arid Zimbabwe: A review. Soil and Tillage Research. 2015;**145**:184-197. DOI: 10.1016/j.still.2014.09.003

[17] Dokyi E, Anang BT, Owusu V. Impacts of improved seed maize technology adoption on productivity and technical efficiency in Northern Ghana. Open Economics. 2021;**4**(1):118-132

[18] Effland A, Saavoss M, Capehart T, McBride D, Boline A. USDA ERS – Innovations in seed and farming technologies drive productivity gains and costs on corn farms. Economic Research Service U.S. Department of Agriculture; 2022

[19] Soane BD, Ball BC, Arvidsson J, Basch G, Moreno F, Roger-Estrade J. No-till in northern, western and South-Western Europe: A review of problems and opportunities for crop production and the environment. Soil and Tillage Research. 2012;**118**:66-87. DOI: 10.1016/j.still.2011.10.015

[20] Pooja VS, Pradhan SS, Kumar P, Singh A. A review on crop establishment methods on weed, yield and quality of maize crop. The Pharma Innovation Journal. 2022;**11**(5):696-698

[21] Saad AA, Das TK, Rana DS, Sharma AR, Bhattacharyya R, Lal K. Energy auditing of a maize–wheat–greengram cropping system under conventional and conservation agriculture in irrigated north-western Indo-Gangetic Plains. Energy. 2016;**116**:293-305. DOI: 10.1016/j.energy.2016.09.115

[22] Rhezali A, Aissaoui AE. Feasibility study of using absolute SPAD values for standardized evaluation of corn nitrogen status. Nitrogen. 2021;**2**(3):298-307. DOI: 10.3390/nitrogen2030020

[23] Rosenfield GH, Guindin-Garcia NO, Ramírez-Aviles L. Satellite remote sensing for maize crop monitoring: A review. Remote Sensing. 2019;**11**(4):431

[24] Chen W, Xu J, Huang W, Zhang L, Li M, Zhang S. A variable rate nitrogen application system for maize using drone and wireless sensor network. Computers and Electronics in Agriculture. 2021;**183**:106045

[25] Smith GS, Chow TE, Gerik TJ. Variable rate nitrogen application in irrigated maize using an unmanned aerial vehicle. Computers and Electronics in Agriculture. 2022;**196**:106236

[26] Singh K, Jain S, Andhra V, Sharma S. IoT based approach for smart irrigation system suited to multiple crop cultivation. International Journal of Engineering Research and Technology. 2019;**12**(3):357-363

[27] Xu J, Chen W, Li M, Huang W, Zhang S, Li L. A UAV-based high-throughput phenotyping platform for maize (Zea mays L.) phenotypic traits. Plant Methods. 2022;**18**(1):5

[28] Chen S, Zhen X, Liu Q, Li X. Multi-source remote sensing data and

UAV-based method for monitoring crop growth. IEEE Transactions on Geoscience and Remote Sensing. 2018;**56**(11):6599-6611

[29] Torres-Sánchez J, López-Granados F, Peña JM. An automatic object-based method for optimal thresholding in UAV images applied to agriculture. Remote Sensing. 2019;**11**(9):1016

[30] Ruiz DP, López-Granados F, Serrano N. Maize yield prediction using Sentinel-2 satellite imagery and machine learning algorithms. Remote Sensing. 2020;**12**(13):2104

[31] Wang X, Zhang B, Li X, Du L, Yang G, Cao W. Integrating UAV-based remote sensing and crop model for maize yield prediction at regional scale. Remote Sensing. 2021;**13**(7):1279

[32] Wang Y, Li F. Maize yield estimation combining machine learning and crop simulation model using multi-source data from UAV and weather station. Computers and Electronics in Agriculture. 2021;**187**:106235

[33] Sharifi A. Using Sentinel-2 data to predict nitrogen uptake in maize crop. IEEE Journal of Selected Topics in Applied Earth Observations and Remote Sensing. 2020. pp. 1-1. DOI: 10.1109/JSTARS.2020.2998638

[34] Paucar LG, Diaz AR, Viani F, Robol F, Polo A, Massa A. Decision support for smart irrigation by means of wireless distributed sensors. In: 2015 IEEE 15th Mediterranean microwave symposium (MMS). 2015. DOI: 10.1109/mms.2015.7375469

[35] Murali K, Sridhar B. A smart agriculture irrigation system using sensor array based IOT. Journal of Physics: Conference Series. 2021;**2062**:012010

[36] Ghosh S, Sayyed S, Wani K, Mhatre M, Hyder Ali H. Smart Irrigation: A Smart Drip Irrigation System Using Cloud, Android and Data Mining. 2016. pp. 236-239. DOI: 10.1109/ICAECCT.2016.7942589

[37] Abagissa A, Behura A, Pani S. IoT Based Smart Agricultural Device Controlling System. 2018. pp. 26-30. DOI: 10.1109/ICICCT.2018.8473126

[38] Munir M, Sarwar I, Ashraf A, Anwar D, Rashid R. Intelligent and smart irrigation system using edge computing and IoT. Complexity. 2021;**2021**:1-16. DOI: 10.1155/2021/6691571

[39] Suma V. Internet-of-Things (IoT) based smart agriculture in India – An overview. Journal of ISMAC. 2021;**3**:1-15. DOI: 10.36548/jismac.2021.1.001

[40] Musa SFPD, Basir KH. Smart farming: Towards a sustainable Agri-food system. British Food Journal. 2021;**123**(9):3085-3099. DOI: 10.1108/BFJ-03-2021-0325

[41] Zia H, Rehman A, Harris NR, Fatima S, Khurram M. An experimental comparison of IoT-based and traditional irrigation scheduling on a flood-irrigated subtropical lemon farm. Sensors. 2021;**21**:4175. DOI: 10.3390/s21124175

[42] Shiferaw B, Boddupalli MP, Hellin J, Bänziger M. Crops that feed the world 6. Past successes and future challenges to the role played by maize in global food security. Food Security. 2011;**3**:307-327

[43] Adamchuk VI, Hummel JW, Morgan MT, Upadhyaya SK. On-the-go soil sensors for precision agriculture. Computers and Electronics in Agriculture. 2004;**44**:71-91

[44] Sharifnasab H, Mahrokh A, Dehghanisanij H, Łazuka E,

Łagód G, Karami H. Evaluating the use of intelligent irrigation systems based on the IoT in grain corn irrigation. Water. 2023;**15**:1394. DOI: 10.3390/w15071394

[45] Kumar VS, Singh CD, Ramana Rao KV, Kumar M, Rajwade YA, Babu B, et al. Evaluation of IoT based smart drip irrigation and ETc based system for sweet corn. Smart Agricultural Technology. 2023;**5**:100248. DOI: 10.1016/j.atech.2023.100248

[46] Asiimwe G, Jaafar H, Asce AM, Haidar M, Mourad R. Soil moisture or ET-based smart irrigation scheduling: A comparison for sweet corn with sap flow measurements. Journal of Irrigation and Drainage Engineering. 2022;**148**(6):04022017. DOI: 10.1061/(ASCE)IR.1943-4774.0001668

[47] Manida M, Ganeshan MK. New agriculture technology in modern farming. International Journal of Management Research and Social Science (IJMRSS). 2021;**8**(3)

[48] Tantalaki N, Souravlas S, Roumeliotis M. Data-driven decision making in precision agriculture: The rise of big data in agricultural systems. Journal of Agricultural Food Information. 2019;**20**(4):344-380

[49] Noriega-Navarrete JL, Salazar-Moreno R, López-Cruz IL. Corn growth and yield models under climate change scenarios. Revista mexicana de ciencias agrícolas. 2021;**12**(1):127-140

[50] Chisanga C. Modelling climate change impacts on maize. CAB Reviews: Perspectives in Agriculture, Veterinary Science, Nutrition and Natural Resources. 2022;**17**. DOI: 10.1079/cabireviews202217008

[51] Abedinpour M, Sarangi A, Rajput TBS, Singh M, Pathak H, Ahmad T. Performance evaluation of AquaCrop model for maize crop in a semi-arid environment. Agricultural Water Management. 2012;**110**:55-66

[52] Sandhu R, Irmak S. Performance of AquaCrop model in simulating maize growth, yield, and evapotranspiration under rainfed, limited and full irrigation. Agricultural Water Management. 2019;**223**:105687

[53] Eshete DG, Sinshaw BG, Gizaw HD, Zerihun BA. Simulating maize water productivity at deficit irrigated field in north West Ethiopia. Sustainable Water Resources Management. 2022;**8**(6):186

[54] Chan K, Millinger MA, Schneider U, Thran D. How diet portfolio shifts combined with land-based climate change mitigation strategies could reduce climate burdens in Germany. Journal of Cleaner Production. 2022;**376**(134200):1-11. DOI: 10.1016/j.jclepro.2022.134200

[55] Faloye OT, Ajayi AE, Alatise MO, Ewulo BS, Horn R. Maize growth and yield modelling using aquacrop under deficit irrigation with sole and combined application of biochar and inorganic fertiliser. Journal of Soil Science and Plant Nutrition. 2020;**20**:2440-2453

[56] Chitsiko RJ, Mutanga O, Dube T, Kutywayo D. Review of the current models and approaches used for maize crop yield forecasting in sub-Saharan Africa, and their potential use in early warning systems. Physics and Chemistry of the Earth, Parts A/B/C. 2022;**2022**:103199

[57] Holka M, Bieńkowski J. Carbon footprint and life-cycle costs of maize production in conventional and non-inversion tillage systems. Agronomy. 2020;**10**(12):1877

[58] Sun SK, Wu PT, Wang YB, et al. Temporal variability of water footprint

for maize production: The case of Beijing from 1978 to 2008. Water Resources Management. 2013;**27**:2447-2463. DOI: 10.1007/s11269-013-0296-1

[59] Liu W, Antonelli M, Liu X, Yang H. Towards improvement of grey water footprint assessment: With an illustration for global maize cultivation. Journal of Cleaner Production. 2017;**147**:1-9. DOI: 10.1016/j.jclepro.2017.01.072

[60] Borsato E, Galindo A, Tarolli P, Sartori L, Marinello F. Evaluation of the grey water footprint comparing the indirect effects of different agricultural practices. Sustainability. 2018;**10**(11):3992. DOI: 10.3390/su10113992

[61] Yegbemey RN, Yegbemey EO, Yabi JA. Sustainability analysis of observed climate change adaptation strategies in maize farming in Benin, West Africa. Outlook on Agriculture. 2017;**46**(1):20-27. DOI: 10.1177/0030727016689638

[62] Cairns JE, Chamberlin J, Rutsaert P, Voss RC, Ndhlela T, Magorokosho C. Challenges for sustainable maize production of smallholder farmers in sub-Saharan Africa. Journal of Cereal Science. 2021;**101**:103274. DOI: 10.1016/j.jcs.2021.103274

[63] Awaad H, Negm A, Abu-hashim M. Introduction to "Mitigating Environmental Stresses for Agricultural Sustainability in Egypt". 2021. pp. 3-14. DOI: 10.1007/978-3-030-64323-2_1

Section 3

Scaling Mechanization and Profitability in Maize Cultivation

Chapter 5

Scaling Mechanization and Profitability in Maize Cultivation through Innovative Maize Planters along with Agroforestry Approach: Sustainable and Climate Smart Approach to Diversify Rice Based Cereal Systems in Various Regions

Rupinder Chandel, Mahesh Kumar Narang and Surinder Singh Thakur

Abstract

Keeping in view declining water tables in India and across the world, low greenhouse gas (GHG) emission and global warming potential (GWP) for maize as compared to rice a study was done on maize planters along with agro forestry concept. The yield for inclined and vertical plate mechanism ranged between 4.96–7.71 $t.ha^{-1}$ and 6.75–8.61 $t.ha^{-1}$, respectively. The increase in maize yield in raised bed planters varied between 0.48–2.57 $t.ha^{-1}$. The maximum yield was recorded from pneumatic raised bed planter with bed of 150 mm height and 711 mm top width (2 rows on each bed). The saving of irrigation water ranged between 9.68–23.69% for raised bed planting (150–290 mm) as compared to flat planting. The specific energy was found minimum for pneumatic raised bed and flat planter as 7.02 and 7.38 $MJ.kg^{-1}$. The energy productivity was found maximum for pneumatic raised and flat planter as 0.14 $Kg.MJ^{-1}$ (cost \$12.60 per ha and \$9.33 per ha) followed by raised bed inclined plate planter as 0.13 $Kg.MJ^{-1}$ and were found economical as compared with ridger+manual sowing method (cost \$77.62 per ha).

Keywords: energy, maize crop planter, water savings, raised bed, pneumatic, maize yield

1. Introduction

Maize due to its various uses in feed (61%), industry (22%) and food sectors (17%), is considered as an internationally important commodity driving world agriculture.

Globally, it is grown in 193.7 million hectare across 170 countries (**Figure 1**), with total production of 1147.7 million metric tonne and average productivity of 5.75 t ha^{-1}. It has attained a position of industrial crop globally as 83% of its production in the world is used in feed, starch and bio-fuel industries [1, 2]. It has emerged as the most cultivated grain in the world, surpassing rice and wheat in 1996 and 1997, respectively [3]. Largest grain crop in India, after rice and wheat is Maize (*Zea mays* L.). It is cultivated in an area of 9.09 million hectares (M ha), with an annual production of 24.26 million metric tonnes (MMT), and an average national productivity of 2.56 metric tonnes per ha (t ha^{-1}) [4]. In US and China are the leading producer accounting for about 38% and 23% respectively and India contributes around 2% of this production chart (26 million MT) in 2016–2017. In the Indian context it generates employment for more than 650 million person-days at farming and the businesses related to it. States such as Karnataka, Rajasthan, Andhra Pradesh and Madhya Pradesh, Bihar contribute towards almost 2/3rd of the national maize production [5]. It is grown in India during rainy (*kharif*), winter (*rabi*) and spring seasons, but major production is in the rainy season [6]. Area under Rabi Maize (>400 thousand ha) is larger than that under *Kharif* maize (>230 thousand ha) in Bihar due to low infestation of insect, pest and diseases as well as slow growth of weeds [7]. The abiotic and biotic stresses listed in descending order of importance are: caterpillars, water stress, stem borers, weevils, zinc deficiency, rust, seed/seedling blight, cutworm, leaf blight and technological parameters. A potential solution for organic maize is to apply the biological control agent Trichogramma strips at around 10 and 17 days crop (100–125 no ha^{-1}; size 5 × 1.50 cm). A study revealed that that by application of Trichogramma pretiosum, 79.2% of egg masses were parasited and maize yield increased by (701 kg ha^{-1}) 19.4% [8].

Water stress during the growing season can decrease grain yields [9]. The FIRB technique save the resources like water, nutrients and labour and also facilitates the greater diversification of the rice-wheat cropping systems and improve the physical properties of soil [10]. The raised-bed planting may enhance maize productivity in part by increasing availability of essential crop nutrients by stimulating microbial activity. Raised-bed planting yielded mean saccharase, urease, protease and phosphatase activities across sampling times in 2006 of 2.3 mg glucose $g^{-1} h^{-1}$, 0.8 mg NH_3–N $g^{-1} h^{-1}$, 10.5 mg glycine $kg^{-1} h^{-1}$, and 0.4 mg nitrophenol $g^{-1} h^{-1}$, 6, 18, 34, and 31% higher than those in flat planting, respectively [11]. It was reported that wide (180 cm) beds produced higher wheat (15%) and maize (26%) yields whereas narrow (65 cm) and medium (130 cm) width beds produced higher maize yields (10%) while wheat yields were only marginally (<5%) higher than the basin treatment. The narrow beds used 3–7% while the medium and wide beds used 16–17% and 18–22% less water than the basins [12]. A 3–4 inch bed height is necessary for maintaining

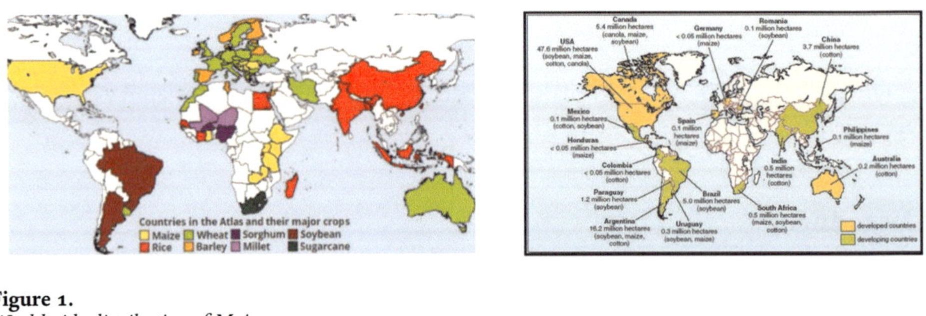

Figure 1.
Worldwide distribution of Major crops.

maximum yield for both corn and soybeans [13]. There was water saving of about 20.4% for wheat crop (for wide beds (107 cm furrow centre gap) and about 16.5% for narrow beds (37 cm furrow centre gap) with grain yield increase of about 13.5% (5.13 and 4.44 t ha^{-1}) and 11.8% (4.33 and 3.82 t ha^{-1}) for maize crop with precision land leveling and raised bed planting compared to traditional land leveling with flat beds planting [14]. Increasing the compost from 5 to 10 ton ha^{-1} increased the yield, protein and K contents in maize crop. The interaction between compost manure (10 ton ha^{-1}) and nano-potassium (500 cm^3 ha^{-1}) or humic acid (10 ton ha^{-1}) recorded the highest mean values for all parameters during both harvest seasons [15]. A study was done for maize (PMH-1) grain in the moisture range of 10–18% wet basis (w.b.), the length of wetted grain increased from 10.01 to 10.65 mm, width increased from 8.57 to 8.70 mm, thickness ranged from 4.63 to 4.97 mm and the angle of repose varied from 23.36° to 28.55° [16] and hue angle (z%) decreased from 14.59 to 14.06 [17]. Maize sown on ridges resulted in greater seed emergence of 89%, plant height of 155.1 cm, and greater grain yield of 6.35 t ha^{-1} [18]. The manual punch planter recorded 61–64% singles, 17–19% multiples, 17–22% missing for speed ranging between 0.5–0.7 km h^{-1} and for soil with 69% clay, 16% silt and 15% sand [19]. A punch planter for corn was evaluated for no-till conditions at the vertical position with 2.5 kPa of vacuum and at a 22° incline with 4.0 kPa of vacuum. Only small changes occurred in the seed meter performance when speed varied from 1 to 3 m s^{-1} [20]. The best seed spacing uniformity and seed emergence ratio were obtained with the no-till planter, and the best seed depth uniformity was obtained with the precision vacuum planter. As forward speed increased, mean emergence time decreased ($p < 0.05$) [21]. The time required to plant one hectare of farmland with manual planter was determined as 3.7 hours [22]. A small maize planter with an independent driving wheel and stationary firming wheels was specially designed and was found suitable as compared to ordinary seeders for complex terrain and heavy soil surface condition [23]. The data showed that planter performance in terms of emergence and plant spacing coefficient of variation (CV) was comparable for most of the meter speeds (17.4–33.5 rpm) among the two seed meters (variable depth and variable seed rate) utilized in the study [24]. For common grain drills, a CV of 20% is an acceptable accuracy achieved by mechanical and pneumatic machines when they are performing well [25]. Panning et al. [26] evaluated a vacuum metering general purpose planter designed for shallow planting of small seeds for sugar beet crop. The most uniform seed spacing occurred at the lowest speed of 3.2 km h^{-1} and decreased as the forward speed increased from 3.2 to 8.0 km h^{-1}. The seed spacing uniformity was not affected by planter forward speed between 4.8 and 11.2 km h^{-1} [27]. A population of 90,000 plants ha^{-1} had the highest grain yields than lower populations for adequate nutrients and water supply. When density/population of plants increases, stalk lodging will increase due to smaller stalk diameter and a slight gain in grain test weight was observed [28–30]. It was reported that as plant population increased, the yield and kernel numbers increased but weight of kernels decreased [9, 31]. Yield reductions from uneven plant distributions ranged from 0 to 31% and averaged 10% [32]. The part of sowing depth real-time control included the module of collect pressure information and the module of sowing depth adjustment and the part of precise control of the sowing spacing included the module of speed acquisition and sowing motor control in a developed intelligent detection and control system for corn precision planter [33].

In a tillage study soil conditions induced fall moldboard plow, spring disk, and no-till were measured and the effects of tillage-induced soil conditions on planting depth, seedling emergence, and early growth of four maize hybrids grown continuously were evaluated on a poorly drained, moderately permeable soils. Surface residue cover averaged 10, 39, and 68% for the moldboard-plow, spring-disk, and no-till tillage systems, respectively. The study revealed that the residue from the previous maize crop remaining on the soil surface had a greater effect on plant growth than did the other soil physical properties measured. Seed placement was shallower and more variable on tillage systems with greater surface residue cover and early growth was delayed by systems with a large percentage of surface residue cover. Tillage systems with the best early growth tended to have the greatest yield, however, yields of hybrids were not always correlated with early growth The increase in seed depth with increasing amounts of tillage may result from decreasing soil strength or from decreasing surface residue cover. The final emerged plant population was least for the no-till system. Populations were similar in the spring disk and fall moldboard plow systems. Populations may have been reduced in the no-till system because of seed decay before germination or because seed was planted near residue pressed into the soil by the planter. Residue near the seed could reduce soil-seed contact and produce an allelopathic effect that can stunt or prevent early seedling growth [34, 35]. Compared with strip-rotating maize no-tillage planter, the maize no-till planter could not only seed and fertilize at the suitable depths, but also decrease soil disturbance and fuel consumption by 69.7% and 19.3%, respectively [36].

Field test shows that the planter has a good performance of trafficability with the ratio of sheering off corn stubble 85% and anti-blocking capacity, thus to finish wheat and maize no-till planting. The variation coefficient of seed depth was 19.4% and 23.4% for wheat and maize, respectively [37]. A rotary drum-type anti-blocking mechanism was developed and mounted in front of each opener shank of the maize planter and the drum was rotated driven by ground wheel at a certain speed. The result showed that the speed ratio was the most significant factor that affecting anti-blocking performance. Based on the results of simulation, the speed ratio of 1.24, the drum diameter of 150mm and 5 bars were the optimum parameters [38]. Ultra high precision placement of seed was also established. Mechanisms that ensure that the seeds planted has zero ground velocity [39].

Apart from planting/sowing technique, the crop selection and rotation, tillage practices have a significant effect on GHG emissions and resource conservation. The 24.8% of global greenhouse gases (GHGs) are emitted by "Agriculture, Forestry and Other Land Use (AFOLU)", including 0.5 Gt carbon dioxide equivalents (CO_2e) yr^{-1} from enteric fermentation and 1.2 Gt CO_2e yr^{-1} from agricultural soils [40]. The principal emissions from agricultural practices consist of (1) methane (CH_4) from enteric fermentation, (2) carbon dioxide (CO_2) from decomposition of soil organic carbon (SOC), and (3) nitrous oxide (N_2O) from synthetic fertilizer and manure [40]. The global warming potential (GWP) of each gas differs, however, with CO_2e values of 34, 3.7, and 298 for CH_4, SOC, and N_2O, respectively [41] (Intergovernmental Panel on Climate Change).

Results show that the GWP of CH_4 and N_2O emissions from rice (3757 kg CO_2 eq ha^{-1} $season^{-1}$) was higher than wheat (662 kg CO_2 eq ha^{-1} $season^{-1}$) and maize (1399 kg CO_2 eq ha^{-1} $season^{-1}$). The yield-scaled GWP of rice was about four times higher (657 kg CO_2 eq Mg^{-1}) than wheat (166 kg CO_2 eq Mg^{-1}) and maize (185 kg CO_2 eq Mg^{-1}), suggesting greater mitigation opportunities for rice systems [42]. Intermittent irrigation in rice reduced methane emissions by 40% whereas application of farmyard manure in rice increased the GWP by 41% [43]. However, practice of mid-season drainage has reduced green house gases equivalent to 270 million tonnes of carbon dioxide and increased the release of nitrous oxide, by about 20,000 tonnes for the same period [44]. It was estimated that CH_4 emissions from global rice fields varied from 18.3 ± 0.1 Tg CH_4/yr (Avg. ±1 SD) under intermittent irrigation to 38.8 ± 1.0 Tg CH_4/yr under continuous flooding [45]. Around 30% and 11% of global agricultural CH_4 and N_2O, respectively, emitted from rice fields and A recent study based on the database from different states in India documented national CH_4 budget estimate of 4.09 ± 1.19 Tg year^{-1} [46]. Open-burning of straw residues also contributes to global warming through emissions of greenhouse gases (GHGs) such as carbon dioxide (CO_2), methane (CH_4), and nitrous oxide (N_2O) [41, 47, 48]. The carbon (C) and nitrogen (N) in the burned straw are emitted as CO_2—C (57–81%), CO—C (5–9%), CH_4—C (0.43–0.90%), and N_2O—N (1.16–1.50%) [49]. The global warming potential for CO_2 is 1 (100 years period), CH_4 is 27–30 (12 years in atmosphere), N_20 is 273 (109 years). Another potent green house gas carbon monoxide reacts with hydroxyl (OH) radicals in the atmosphere, reducing their abundance. As OH radicals help to reduce the lifetimes of strong greenhouse gases, like methane, carbon monoxide indirectly increases the global warming potential of these gases [50]. This means that a methane emission is projected to have 28 times the impact on temperature of a carbon dioxide emissions of the same mass over the following 100 years assuming no change in the rates of carbon sequestration. More than half of the South Asian population's livelihood capabilities are at risk due to rising temperature, droughts, erratic and isolated rainfalls, floods resulting in decline of crop yield, water logging/water scarcity, reduced farm income and migration [51]. Growing rice in rotation with soybean and planting hybrid cultivars, drainage twice may result in reduced CH_4 emissions. However, mineral-soil dressing on peat could have a significant impact on suppression of CH_4 emissions from beneath the peat reservoir [52, 53]. The study suggests that adoption of rice-rice-rape (*Brassica napus* L.) cropping system would be beneficial for greenhouse gas emission mitigation and as good cropping pattern in double rice cropped regions [54]. The monoculture in any cropping system causes more insect-pest attack, depletion of soil organic carbon, underground water, more use of fertilizers etc. Therefore crops should be grown in rotation specially with legumes to maintain soil health, reduce use of fertilizers, break insect-pest cycle thereby reducing use of chemicals, pesticides etc. A study revealed that crop residue return might be most effective in increasing crop yields and WUE in corn crops with a tillage depth > 20 cm, for cold conditions (<10°C), moderate nitrogen fertilization (0–150 kg ha^{-1}), growth of a single crop per year and high soil organic matter content (>15 g kg^{-1}) [55]. By assuming, the crops which had C:N ratio more than the threshold C:N ratio (50) and plant biomass higher than the threshold biomass (25 g/plant) were considered as having higher carbon sequestration potential. Based on these, the carbon sequestration potential of maize, sorghum and pearl millet was higher as compared to rice, finger millet and soybean [56]. Croplands worldwide and specially in intensively cultivated regions such as North America, Europe, India and intensively cultivated areas in Africa, such as Ethiopia could sequester between 0.90 and 1.85 Pg C/yr, i.e. 26–53% of the target of the "4p1000 Initiative: Soils for Food Security and Climate". Soil carbon sequestration and the conservation of existing soil carbon stocks is an important mitigation pathway to achieve the less than 2°C global target of the Paris Climate Agreement [57]. The crop water productivity for maize (1.80 kg m^{-3}) is higher as compared to wheat (1.09 kg m^{-3}), rice (1.09 kg m^{-3}), cotton$_{seed}$ (0.65 kg m^{-3}), cotton$_{lint}$ (0.23 kg m^{-3}) [58]. The carbon dioxide sequestration potential of corn is 20 tonne ha^{-1}. Depending upon location and the specific management practices implemented, the Climate Exchange bases Michigan carbon payments on approximately 1.0–1.5 tons of carbon dioxide equivalent per ha per year [59]. Soil biota includes earthworms, nematodes, protozoa, fungi, bacteria and different arthropods. Detritus (plant leaves, roots, stubble mulch etc.) resulting from plant senescence (final stage of plant growth) is the major source of soil carbon and above micro organisms decomposes these materials to help maintain nutrient cycling and organic carbon in soil. The organic matter content, especially the more stable humus, increases the capacity to store water and store (sequester) C from the atmosphere. The fastest way to gain soil carbon is to convert to long term no till, adding high carbon crops (corn and wheat) and adding cover crop mixture high in carbon (grasses primarily but also legumes to stabilize soil carbon). Along with GHG emissions, the depletion of ground water table under the existing 'Rice-Wheat' rotation in the erstwhile food bowl (Indo-Gangetic Plains) of the country has also alerted the state governments to diversify the cropping system and maize is a promising substitute. The wheat and paddy requires respectively 3–4, 30–35 irrigations per crop cycle where as maize crop requires 8–15 irrigations (depending upon rainfall) per crop cycle (each irrigation 50 mm). However, national productivity of maize is considerably lower than the global standards and there lies immense scope for improvement in farming technologies. Thus planters especially raised bed planters play a crucial role in achieving optimum maize crop stand, plant spacing, planting depth and higher yields in a sustainable way. Therefore, the feasible low cost flat and raised bed row crop precision planters were evaluated for sowing of maize crop and yield, energetic, irrigations aspects were studied.

2. Material and methods

2.1 Experimental site and maize variety description

The two raised bed planters with inclined and vertical metering plate mechanism, manual planter, flat inclined plate planter and pneumatic raised bed/flat planters were used for this study and the manual sowing on ridges was taken as control plot. The various field and crop parameters are shown in **Table 1**. The experiments were conducted at Department Farm Machinery and Power

Parameters	Detail/Value
Soil type	Sandy loam (2014–2017)
Soil type	Arid brown soil (2017–2020)
Longitude	75°49″09.082″ E, 75.4216702° E
Latitude	30°54′39.286″ N, 31.1797347° N
Mean monthly rainfall, mm	130.88
Mean maize seed characteristics L, B, T, mm	9.0, 7.8, 5.6
Degree of sphericity	0.0670452846592
1000 seed weight, g	286.0
Angle of repose	27.64

Table 1.
Soil type and location, mean rainfall for experimental areas and maize seed parameters.

Engineering Research Farm and farmer's fields during 2014–2020. The field was prepared with mould board plough, two operations of disc harrow followed by two operations of cultivator and one operation of planker and laser leveler to get the best sowing uniformity, the most uniform sowing depth, and maximum emergence percentage with various planters [60]. The various specifications of planters used for this study are shown in **Tables 2** and **3**. The maize crop was sown with various planters during 2014–2020 and sowing was done in East-West direction and North South directions. The seed rate was kept as 20 kg ha^{-1} and 275 kg ha^{-1} urea and 137.5 kg ha^{-1} DAP (N, P & K as 125, 60, 30 kg ha^{-1}), 50 kg ha^{-1} muriate of potash was applied. The weeds were controlled by chemical attrazine which was applied for weeds control (within 10 days of sowing) @2000 g ha^{-1} in 500 l water and mechanically by tractor operated 3-row sweep type weeder when crop height was 300 mm and during further stages of growth. Urea was applied in three splits 1/3 rd at sowing, 1/3 rd at 300 mm height and rest at silking stage. For fall armyworm insect control 1.0 ml corazen 18.5 S.C. per litre was applied for 20 days crop in 300 l water and later on according to stage of maize crop water upto 500 l was used (per ha area).

Parameters	**Raised bed inclined plate planter P_{rbip}**	**Raised bed vertical plate planter (P_{rbvp})**	**Flat Inclined plate planter P_{fip}**	**Pneumatic raised bed P_{prbvp}/flat planter P_{pfvp}**	**Ridger + manual sowing R_{ms}**
Required tractor power, KW	33.60	33.60	26.11	29.84 (dual clutch)	33.60 + 0.60H. E.*
Size of machine, L × B × H, mm	1670 × 3040 × 1250	1460 × 2050 × 1220	1350 × 2470 × 1065	2032 × 1524 × 1219	1000 × 2000 × 1200
Number of rows	4	4	4	4	2
Furrow opener type	Reversible shovel type	Reversible shovel type	Reversible shovel type	full runner type	Ø
Bed maker	Plough type	Plough type	Ø	Plough type	Plough type
Bed/ridge height/ top width, mm	230/350	150/350	Ø	150/711	290/Ø
Row spacing, mm	675	675	675	675	675, 600
Metering plate(mp) material	Aluminum	Mild steel	Mild steel	SS-304	Ø
Metering plate diameter, mm	160	180	160	215	Ø
Seed metering mechanism	Inclined plate with cells on periphery	Vertical plate with spoons on periphery	Inclined plate with cells on periphery	Vertical plate with holes on periphery	Manual
No. of cells/spoon on each plate	8	12	24	26 (5mm hole)	Ø
Adjacent cell/spoon spacing, mm	55	50	12	—	Ø
mp inclination with horizontal	45°	90°	45°	90°	Ø
Height of seed drop, mm	740	970	940	100	Ø

Parameters	Raised bed inclined plate planter P_{rbip}	Raised bed vertical plate planter (P_{rbvp})	Flat Inclined plate planter P_{fip}	Pneumatic raised bed P_{prbvp}/flat planter P_{pfvp}	Ridger + manual sowing R_{ms}
Ground wheel (gw) diameter, mm	508	400	420	356	Ø
Speed ratio gw:mp and mode of power transmission to metering plate	1.25:1 chain sprocket and bevel gear	1.25:1 chain sprocket	1.25:1 Chain sprocket and bevel gear	Chain and sprocket	Ø
Weight, kg	515	315	250	300	175
Seed covering device	Mild steel Strips	Cast Iron Roller	No device	Fiber plastic wheels/zero pressure pneumatic press wheel	Manual

*H.E.—human energy, KW [61].

Table 2.
Specifications of machines used for raised bed planting of maize crop and their operational parameters.

Particulars	Unit	Energy equivalent MJ $unit^{-1}$	References
Human labor	h	1.96	[62, 63]
Machinery	h	62.70	[63]
Diesel fuel	L	56.31	[62, 64]

Table 3.
Various energy equivalents for input operations and sources.

2.2 Meteorological data

The average minimum and maximum temperatures were 17.59°C and 29.67°C, respectively, whereas the mean temperature was 23.53°C based on meteorological data. The average annual rainfall was 653.84 mm. The mean sunshine duration was 7.43 h, mean number of rainy days recorded was 1.91 and mean wind speed was 3.53 km h^{-1} (**Figure 2**).

2.3 Planter dimensions, specifications and material description

The specifications of various planters are shown in **Table 2**. The inclined plate with cells and vertical plate with spoons/holes were used as metering mechanisms for seed metering. The furrow openers of planters were made of steel alloy hard. The power transmission to metering plate was given through bevel gear for raised bed inclined plate and flat plate planter and was through chain for raised bed vertical plate planter. In case of raised bed vertical plate planter the hopper size was 140 × 300 mm (l × b) and horizontal distance between inner side of plate to outer side of spoon/spoon length/diameter were 35/15/Φ11 respectively. The hopper size for flat plate planter was 100 × 110 mm (1 hole Φ30mm).

DOI: http://dx.doi.org/10.5772/intechopen.111766

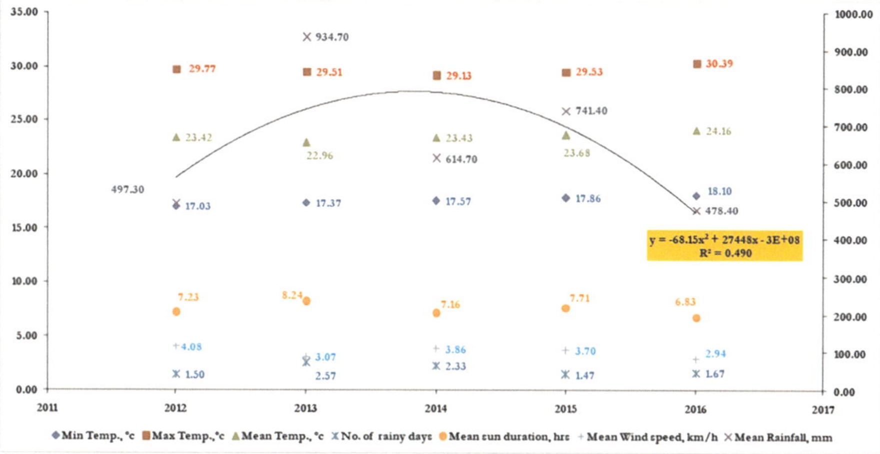

Figure 2.
Mean annual meteorological data from 2012 to 2016.

Runner type opener was used as the fields were well prepared, comprising of sand to loam soils and free from weeds and residue etc. The front section of the opener is 'V'-shaped (in transverse cross section) and extends below the wider rear portion (**Figure 3**). Runner openers are not used in soils with high clay content as the sliding action of the opener causes 'smearing' along the base and walls of the furrow which severely restrict subsequent root development. Similarly reversible shovel type furrow openers causes less disturbance in soil, requires less draft and are easy in construction, cheaper and easily repairable. The seed covering device like mild steel strip (light weight), cast iron roller (for seed covering and bed shaping), zero pressure pneumatic press wheel were used. Zero pressure pneumatic press wheels are continual flexing which makes them self cleaning. The function of covering device in planter is to place the seed in contact with the moist soil, cover them to the proper depth, press the soil firmly around the seeds and leave the soil directly above the row loose enough to minimize crusting and promoting easy emergence. Agitator & sliding orifice type metering mechanism was used for fertilizer metering. Material used for ground wheel, bed former and fertilizer metering mechanism was mild steel. The material used for Pneumatic raised bed/flat planter was mostly Aluminum to make it light weight and cause less compaction of soil. The germination data were recorded for the different rows planted by planters/method and were analyzed for quality, missing, multiples and precision indices. The data was also analyzed statistically. Similarly maize yield and water requirements were also recorded during these experiments. The raised bed inclined plate (**Figure 4**) planter (4-row), raised bed vertical plate planter (2-row) (**Figure 5**) were designed to sow one line of maize on each bed. The mean maize grain length, width and thickness were 9.0, 7.80 and 5.60 mm, respectively. Therefore, cell radius in planter plate was kept as 10 mm and thickness as 8 mm. The angle of repose was 27.64° [16]. Therefore, plate was inclined at 45° i.e. more than angle of repose for free fall of maize seed during field operation. The flat inclined plate planter (**Figure 6**) and pneumatic planter were able to sow 4 rows of maize at a spacing of 675 mm. The ridger was used to make ridges at 600 mm and 675 mm distance and manual plating of maize was done at plant to plant spacing of 200 mm. In case of manual planting around 160 man-h ha^{-1} were involved in sowing operation.

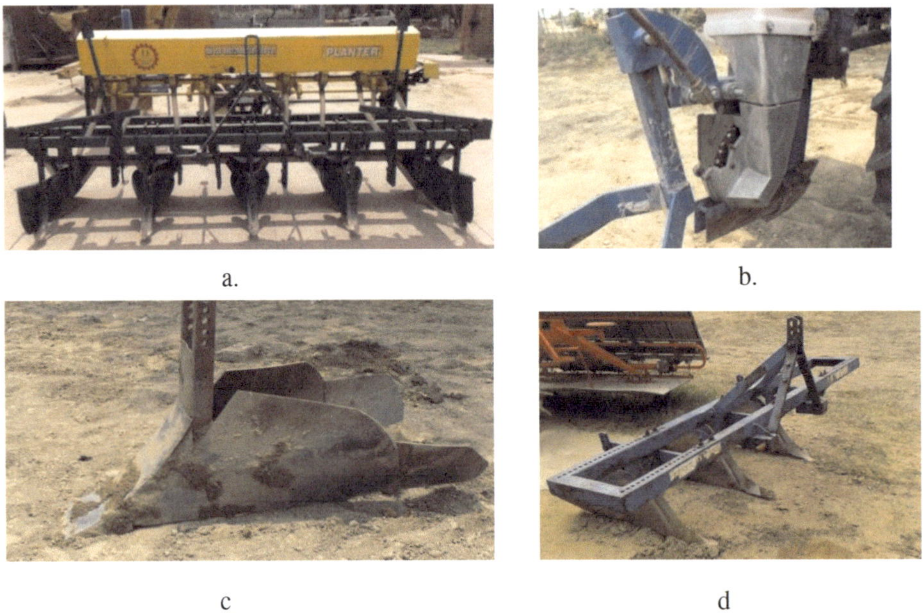

Figure 3.
(a) A view of reversible shovel furrow opener in inclined plate planter, (b) metering plate and full runner opener for pneumatic planter, (c) bed former for pneumatic raised bed planter and (d) ridger.

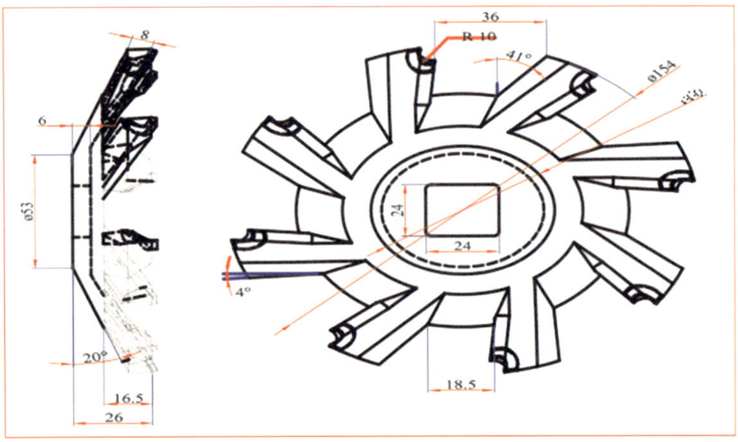

Figure 4.
Line diagram for metering plate of raised plate planter.

The manual planter (**Figure 7**) is cheap, suitable for hilly/plain regions or for land with undulating topography. One persons pulls this machine from front with the help of a rope and other person pushes it from ergonomically designed handle for uniform movement of metering plate and placement of maize seed in soil at proper depth. The physical power output of a male agricultural worker is approximately 75 W and for female worker is 60 W sustained for an 8–10 hours work per day [61]. A furrow opener is provided for opening of soil in manual planter (M_{vp}) and ground wheel is provided for easy movement of planter. The metering plate is driven by it through

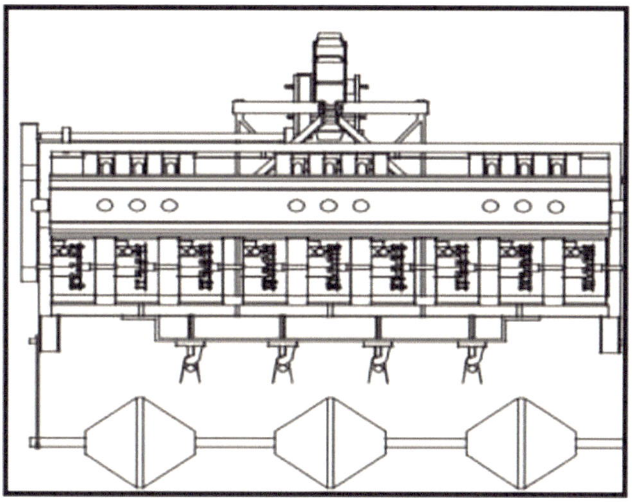

Figure 5.
Raised bed vertical plate planter and bed inclined spoon type vertical metering plate.

Figure 6.
Flat inclined plate planter with U shaped inclined plate planter.

Figure 7.
Manual planter with spoon type vertical metering plate.

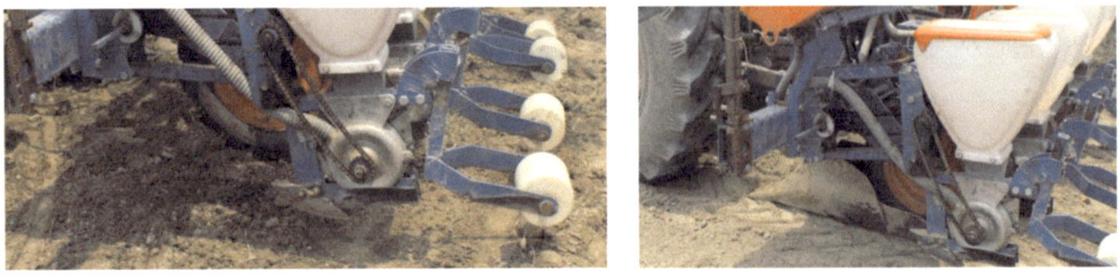

Figure 8.
View of chain transmission system, press wheel and bed maker for low cost Pneumatic precision planter.

Figure 9.
Chain and sprocket mechanism for varying plant to plant spacing.

chain sprocket transmission. In this planter the metering system used is vertical plate with spoons system. The approximate weight of manual maize planter is 20 kg and mean depth of seed placement was 22.35 mm. The view of chain transmission system to metering plate, press wheels, bed former for pneumatic raised bed planter are shown in **Figures 8** and **9**.

Indices for planter performance analysis—All the row crop planters were calibrated in lab for desired seed rate and plant to plant spacing and variation in field was compared with theoretical spacing.

Various indices were used to calculate accuracy of planting of various planters, the description of each index is given below.

2.3.1 Multiple index

Multiple index (D) is the percentage of spacing that are less than or equal to half of the theoretical spacing. D, is an indicator of more than one seed dropped within a desired spacing:

$$\mathrm{D} = N/n_1 \tag{1}$$

where N = total number of observations and n_1 = number of spacing's in the region less than or equal to 0.5 times of the theoretical spacing.

2.3.2 Quality of feed index

It is the percentage of spacing that are more than half, but not more than 1.5 times the theoretical spacing. Quality of feed index, A, is the measure of how often the seed spacing were close to the theoretical spacing [65]. The quality of feed index is mathematically expressed as follows:

$$A = N/n_2 \tag{2}$$

where N = total number of observations and n_2 = number of spacing's between 0.5 times the theoretical spacing and 1.5 times of the theoretical spacing.

2.3.3 Miss index

It is the percentage of more than 1.5 times the theoretical spacing. Miss index, M, is an indicator of how often a seed skips the desired spacing and expressed as:

$$M = N/n_3 \tag{3}$$

where N = total number of observations and n_3 = number of spacing in the region more than 1.5 times of the theoretical spacing.

2.3.4 Precision Index

Precision Index, C, is a measure of the variability in spacing after accounting for variability due to both multiples and skips. The degree of variation is the coefficient of variation of the spacing that are classified as singles, and expressed as:

$$C = ref\ X/S_2 \tag{4}$$

where, S_2 = sample standard deviation of the $n2$ observations and $Xref$ = Theoretical spacing.

Energy input calculations—The various energy equivalents are shown in **Table 3** and energy indices were calculated using following formulae.

- Fuel energy consumption $\text{MJ ha}^{-1} = \text{fuel consumption } (\text{l h}^{-1}) \times \text{fuel energy equivalents } (\text{MJ l}^{-1})/\text{effective field capacity } (\text{ha h}^{-1})$ (5)

- Human energy consumption $\text{MJ ha}^{-1} = \text{no. of human labour used} \times \text{time (h)} \times \text{human energy equivalent } (\text{MJ h}^{-1})/\text{area covered (ha)}$ (6)

- Energy embodied in machinery $\text{MJ ha}^{-1} = \text{weight of specific machine (kg)} \times \text{energy equivalent of machinery } (\text{MJ kg}^{-1})/\text{wear out life of machine (h)} \times \text{effective field capacity } (\text{ha h}^{-1})$ (7)

The energy involved in various planters, mechanical weeders, field preparation, combine harvester, biocides, fertilizer, electricity was considered for energy calculations. The various forms of direct and indirect energy were also calculated for row crop planters and other sowing methods.

2.4 Analysis and economics

Analysis of maize crop yield was done for various planters. The saving in irrigation water, CO_2 emissions were recorded and compared for all the planters. The economics was calculated for all planters using fixed and variable costs for each planter and energy calculations were also done.

3. Results and discussions

The sowing of maize was done with various planters (**Figures 10–12**) and techniques. The operational parameters were recorded for various planters and shown in **Table 4**. The fuel consumption and field capacity for raised bed inclined plate planter were 7.92 l ha^{-1} and 0.60 ha h^{-1}. The mean standard deviation in spacing was 0.92 cm for raised bed inclined plate planter whereas it was 1.67 cm (+0.75 cm) for raised bed vertical plate planter.

The view of ridge formation for manual sowing is shown in **Figure 12**. The field operational parameters of the various row crop planters/methods are shown in **Table 4**.

The maize sowing with pneumatic raised bed planter and pneumatic flat planter is shown in **Figures 13** and **14** respectively. The emergence of maize crop sown with raised bed inclined plate planter and pneumatic raised bed planter is shown in **Figure 15**.

The field observations revealed that higher missing index was either due to higher speed or the 'U' shaped design of metering plate which lead to stucking of two seeds in one cell. Thus this planter design requires human intervention and more human energy for planting accuracy. The design of plate of raised bed inclined plate planter was like 'open loop' (**Figure 10**) which encountered no stucking of seeds in the field operation.

The various parameters recorded at germination stage are shown in **Table 5** and represented in **Figure 16**.

Standard deviation remains a widely used standard of measure for within-row plant spatial variation, and targets the mechanics of the planter as causative for non-uniformity. The grain yields appeared to increase 110 kg ha^{-1} for every 1 cm decrease in standard deviation and change in yield per 1 cm improvement in plant spacing uniformity ranged from 27 to 152 kg ha^{-1}; respective to location [66]. The correct seed metering unit setup is very critical to obtain expected performance from planting technology [24]. The planters were operated between speed range of 1.87–3.79 km h^{-1}. The low speed of planter minimizes the Intra-row spacing by reducing the creation of skips and multiple-plant hills that cause, more so the latter, barren stalks and reduced grain weight per ear [66, 67]. The lowest standard deviation in spacing was achieved by raised bed inclined plate planter design (0.92 cm), which shall lead to higher yield returns. However quality of feed index

Figure 10.
Maize sowing with raised bed inclined plate planter and view of metering plate.

Figure 11.
Maize sowing with raised bed vertical plate planter.

Figure 12.
Ridge formation with ridger for manual sowing.

Operational parameters	Raised bed inclined plate planter	Raised bed vertical plate planter	Flat inclined plate planter	Pneumatic raised bed planter	Manual flat planter	Ridger + manual sowing	Pneumatic flat planter
F_C, l ha^{-1}/human energy KW	7.92	10.10	9.27	7.50	0.15 KW*	8.03	6.25
S, km h^{-1}	3.24	2.64	3.79	1.87	0.46	2.21(R) +0.11 (MS) 1.16″	2.22
C_e, ha h^{-1}	0.60	0.49	0.48	0.50	0.23	0.0061	0.60
d, mm	40.26	40.10	33.63	35.16	23.33	23.45	35.25

F_C, *fuel consumption;* S, *forward speed;* C_e, *effective field capacity;* d, *depth of seed placement.*
*[61]—"Mean speed of ridger + manual sowing technique".

Table 4.
Field operational parameters for various row crop planters.

Figure 13.
Maize sowing with pneumatic raised bed planter.

Figure 14.
Maize sowing with pneumatic flat planter during 2017.

Figure 15.
Emergence of maize crop sown with raised bed inclined plate planter (left, 1-row/bed) and pneumatic raised bed planter (right, 2-rows/bed).

Parameters	Raised bed Inclined plate planter	Raised bed vertical plate planter	Flat Inclined plate planter	Ridger + manual Sowing		Pneumatic raised bed planter	Pneumatic flat planter	Manual planter	p-value	*f*-value
				60.00	**67.5**					
QFI, %	67.44a	48.04a	41.09a	55.22a	39.29a	87.99a	85.25a	76.85a	0.176*	1.965
MI, %	19.00ab	17.52a	30.28ab	44.78ab	53.57b	7.64ab	10.97a	12.75a	0.024*	4.501
MUI, %	13.63a	34.33b	21.46a	0.00c	0.00c	4.37a	3.79a	10.61a	0.002*	31.438
Intra-row spacing, cm	0.92a	1.67a	1.35a	3.40b	2.14a	1.27a	1.55a	2.36a	0.001*	12.332
C, %	4.63a	8.34bc	6.74ab	16.95d	10.69c	6.35a	7.74a	11.82a	0.003*	61.155
PTP spacing, cm	19.35a	18.69a	19.16a	27.87b	27.10b	18.52a	18.92a	18.24a	0.000*	54.035

**Significant at 5% level.*

Table 5.
Performance of raised bed inclined plate and vertical plate planter based on germination data attributes.

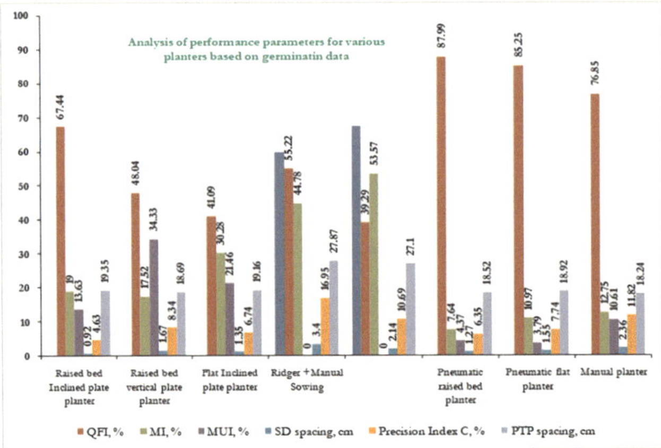

Figure 16.
Graphical representation of various performance parameters for maize planters based on field germination data of maize crop.

was higher for pneumatic raised bed planter (87.99%) and pneumatic flat planter (85.25%). The lowest missing index (7.64%) was recorded for pneumatic raised bed planter and lowest multiple (3.79%) index was observed for pneumatic flat planter. The precision indices for raised bed inclined plate planter, pneumatic raised bed and flat planters were 4.63%, 6.35%, 7.74% respectively. The intra-row spacing for pneumatic raised bed and flat planter were 1.27 cm and 1.55 cm which resulted in higher grain yield. The Intra-row spacing of raised bed vertical plate planter, inclined plate planter were 1.67 cm, 0.92 cm and that of flat planter was 1.35 cm. The forward speed for vertical plate planter was 2.64 km h^{-1} and intra-row spacing was observed as 1.67 cm. The forward speed for raised bed inclined plate planter was 3.24 km h^{-1} and Intra-row spacing was observed as 0.92 cm. The SD increased at faster planting speeds but variation of intra-row spacing with change in forward speed of planter was low in case of inclined plate as compared to vertical plate. Thus sowing with different mechanical planters certainly affected plant population, stand uniformity with mean standard deviation (SD) of within-row plant spacing and consequently maize yield [68]. A view of mechanical weeding operation in raised bed maize crop with sweep type weeder and crop at growing stage are shown in **Figures 17** and **18** respectively. After maturity, maize harvesting was done and yield data was recorded which is shown in **Figures 19** and **20** and represented in **Table 6**.

The yield for ridger + manual sowing method was found more for 60.0 cm spacing (5.38 t ha^{-1}) and lower for 67.5 cm spacing (4.56 t ha^{-1}). The optimum bed design, exposed bed area to sunlight is necessary for better root formation, canopy formation, irrigation water productivity and water drainage.

The maximum cob grain weight of 0.117204 kg (at 10% m.c., w.b.), grain yield of 8.61 t ha^{-1} was observed for pneumatic raised bed planter with number of grains per cob as 410, plant population as 84,095 [9, 69] and 1000 grain weight as 285.86 (at 10% m.c., w.b.), owing to highest QFI as 87.99%, appropriate seeding depth of 35.16 mm and wider row spacing of 67.5 cm appropriate spacing between plants (row spacing and plant to plant spacing) resulted into non overlapping of inter row maize canopies, uniform exposure for all plants to sunlight, higher grain filling and grain weight. The higher yield for pneumatic raised bed planter with 2-rows of maize per bed revealed that 2-rows per bed for 150 mm bed height to be optimum for better crop growth and yield. The difference between QFI for pneumatic raised bed and flat

Figure 17.
Mechanical weeding operation in raised with 3-row sweep type weeder.

Figure 18.
A view of maize crop at growing bed maize crop stage.

planter was found as 2.74% and corresponding yield increase for pneumatic raised bed planter was 3.14% .

The flat inclined plate planter had lower yield of 4.98 t ha^{-1} owing to high missing index and multiple index and low QFI as 41.09%. Due to more multiples, 1000 grain weight per cob was low as 267.40 g because of more competition among plants for nutrients [70]. The grain yield and QFI for raised bed vertical plate planter was 6.75 t ha^{-1}, 48.04% and for manual flat planter was 7.42 t ha^{-1}, 76.85% respectively. In mechanical vertical plate mechanism a slight jerk in field resulted in skip of seeds at various points and more multiples/missings at other points [22]. The difference in QFI for raised bed inclined plate planter and flat inclined plate planter was 26.35% and yield increase for raised bed inclined plate planter was 35.41%. The manual flat planter

Figure 19.
A view of maize crop at maturity stage.

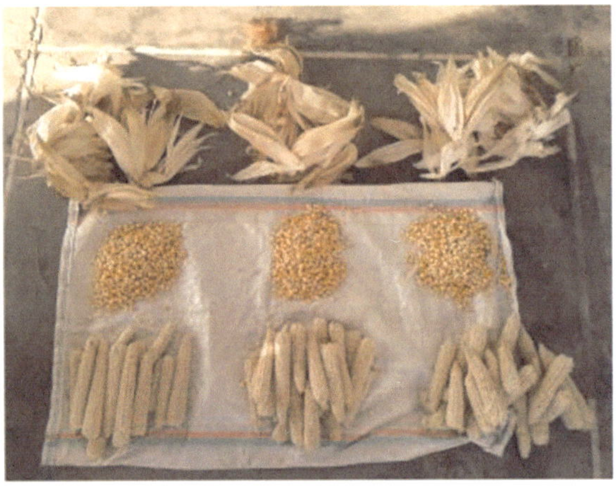

Figure 20.
A view of maize grain samples from various trials.

is economical, easy to operate and suitable for maize planting by small and hilly area farmers [22].

The raised bed inclined plate planter had plant population as 63,623 and cob grain weight as 0.010109 kg. But due to highest precision index 4.63% and more accurate plant to plant spacing 19.35 cm, seed placement at appropriate depth of 40.26 mm, the maize plants sown with this planter recorded maximum 1000 grain weight as 286.57 g and higher grain yield as 7.71 t ha^{-1} [9]. Due to lower missing index, better crop stand and canopy formation it lead to more sunlight exposure and healthy grains with a recorded more maize yield [25, 71].

In case of manual and raised bed vertical plate planter the QFI was higher (76.85%) for manual planter and missings, multiples were higher in raised bed vertical plate planter as 17.52%, 34.33% respectively. The missings may be attributed to higher speed in case of raised bed vertical plate planter (2.64 km h^{-1}) as compared to manual planter (0.46 km h^{-1}). The difference in QFI for manual flat planter and

Maize crop parameters	Raised bed vertical plate planter	Raised bed inclined plate planter	Ridger +manual sowing		Flat Inclined plate planter	Pneumatic raised bed planter with press wheel	Pneumatic flat planter with press wheel	Manual planter
			60.0	67.5				
Varieties	PMH-1, PMH-2	PMH-1, DKC-9108, 1844	PMH-1, DKC-9108	PMH-1	PMH-1, PMH-2	DKC-9108, DKC-9164, Pioneer-1899, 1844	DKC-9108, DKC-9164, Pioneer-1899, 1844	PMH-2, DKC-9108, DKC-9164, Pioneer-1899, 1844
Area, ha	7.0	18.0	8.2	8.2	7.2	195.0	195.0	5.0
Period	07/2014, 07–10/2015	07–10/2015, 6/2016, 02–06/2017, 02/2017	07–10/2015, 02–06/2017, 07–10/2019	07–10/ 2015	07/2014, 07–10/2015, 02–06/2017	07–10/2019, 02/2020	07–10/2019	07/2014, 07–10/2019
Mean ± S.E. p-value 0.001, f-value 10.629	6.75 ± 0.09	7.71 ± 0.17	5.38 ± 0.56	4.56 ± 0.0	4.98 ± 0.51	8.61 ± 0.26	8.34 ± 0.1.07	7.42 ± 0.82
1000 grains weight (m.c. w.b.10%)	275.40	286.57	220.90	249.40	267.40	285.86	282.31	269.40.0
Plants per ha	71,394	63,623	30,581	26,073	68,171	84,095	79,065	73,154
No of grains per cob	359	353	396	395	318	410	402	405
Cob grain weight after moisture correction (at 10% mc w.b.), kg	0.098995	0.10109	0.0795555	0.10709	0.0710895	0.117204	0.113489	0.098745

Table 6.
Results obtained for maize yield (t ha^{-1}) (at 10% m.c., w.b.) sown with various planters during different year experiments.

raised bed vertical plate planter was 28.81% and yield increase for manual planter was 9.03%.

The more height of bed (290 mm) and low depth of sowing in manual method (23.45 mm) lead to lower germination/plant population and lower yield (4.56–5.38 t ha^{-1}). It may be attributed to fact that seed was placed close to soil crust and in low moisture, rapid drying zone and root formation was not appropriate. The difference in maize yield between manual flat planter and manual sowing method (2.04–2.86 t ha^{-1}) also shows the importance of initial soil tillage. However seed metering mechanism in planter is most crucial to obtain optimum plant population, crop stand, growth and yield [24, 72]. The seeding depth for full runner type furrow opener and reversible shovel type furrow opener were 35.16 mm and between 23.33 and 40.26 mm respectively and corresponding plant emergence ranged between 79,065–84,095 and 63,623–73,154 respectively due to low soil resistance. The full runner type furrow opener and reversible shovel type furrow opener were found suitable for sandy loam soil [73, 74]. The depth of seed placement can be attributed to furrow opener type, depth setting, downforce (applied due to weight of planter), pull force, weight of machine, bed maker attachments. The bed maker attachments facilitates tillage in front of furrow opener by cutting, breaking and moving of soil and facilitated deeper placement of seed [34] due to friable condition of soil, which ultimately resulted in maximum plant emergence. The plant population among various planters also showed the benefits of light weight covering device like M.S. strips and zero pressure pneumatic wheels behind the seeding line. The light weight covering device enables furrow closure and seed soil contact for maximum germination and minimal compaction of seeds [75] as low weight covering device leaves the soil in crumbly condition which enables germinated seed to emerge from soil crust with lowest force. The effect of various planting mechanisms (metering, furrow opener and soil covering device), planter speed was found significant on SD value, precision index and maize yield ($p < 0.05$).

The saving in water with raised bed inclined plate planter, raised bed vertical plate planter, ridge planting, pneumatic raised bed planting was 31.25 cm, 15.87, 18.62, 38.85 cm ha^{-1}, respectively as compared to flat planting (**Table 7**). The saving of irrigation water ranged between 9.68 and 23.69% for raised bed planting as compared to flat planting [14]. The CO_2 emissions in kg ha^{-1} for raised bed inclined plate planter, raised bed vertical plate planter, ridge planting and flat planting were found to be 20.91, 26.66, 24.73 and 21.20, respectively and for pneumatic raised bed planter was 19.80 kg ha^{-1}. The maize yield increase were found to be 3.98, 3.39 and 1.33 t ha^{-1} for raised bed inclined plate planter, raised bed vertical plate planter, ridge planting as compared to flat planting. The data collected from sub-mountainous rainfed area revealed that under rainfed conditions (rainfall between 150–950 mm, yearly 944.87 mm, *Kharif* 770.21 mm June-October, *Rabi* 186.89 mm October To February) the maize crop yield lied in between 3.5 and 4.0 t ha^{-1} during *Kharif* season.

It is clear from the graphical representation that the highest irrigation water requirement (656 mm/acre) was for flood irrigation (**Figure 21**). The prediction equation for irrigation water (cm/ha) as a function of height of bed (cm) was obtained as:

$$y = 0.091x^2 – 3.339x + 164.3 \quad (8)$$

The prediction equation for maize yield (t ha^{-1}) as a function of height of bed (cm) and planter design was obtained as:

Planting method	Height of bed	Diesel consumption (l ha^{-1})	Irrigation water (cm ha^{-1})	Cost of operation (Rs/ha)	Yield (t ha^{-1})	1 l diesel equiv. kg CO_2 (g)	CO_2 emissions (kg ha^{-1})	Saving in water (cm ha^{-1})	% saving in water as compared to flat planting	% Yield increase per ha compared to flat planter	Yield increase per ha compared to flat (t ha^{-1})
Raised bed inclined plate planter	230	7.92	132.75	1170.03	7.71	2640	20.91	31.25	19.05	+31.53	+1.98
Raised bed vertical plate planter	150	10.10	148.13	1646.26	6.75	2640	26.66	15.87	9.68	+8.60	+0.54
Ridge planting	290	9.27	145.38	6209.70	4.56	2640	24.73	18.62	11.35	+7.64	+0.48
Flat Planting	0	8.03	164.00	1590.24	4.98	2640	21.20	0.00	0.00	0.00	0.00
Pneumatic raised bed planter	150	7.50	125.15	1022.35	8.61	2640	19.80	38.85	23.69	+40.92	+2.57

Table 7.
Irrigation water requirement, maize yield and CO_2 emissions with various planters.

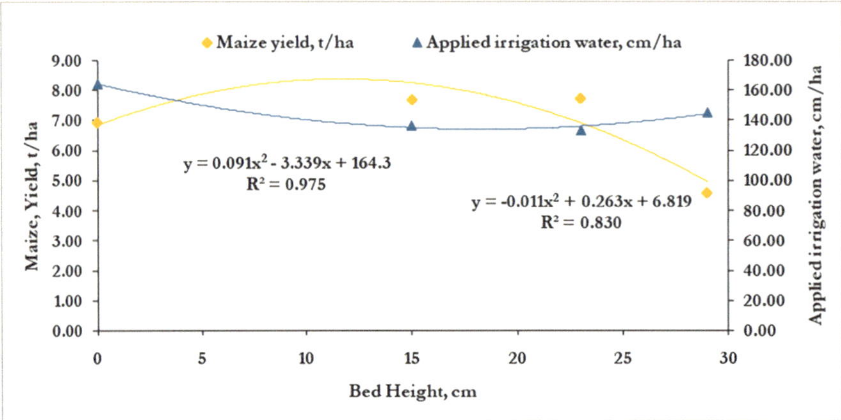

Figure 21.
Maize yield attributed to planter parameters, bed applied irrigation water for various sowing methods.

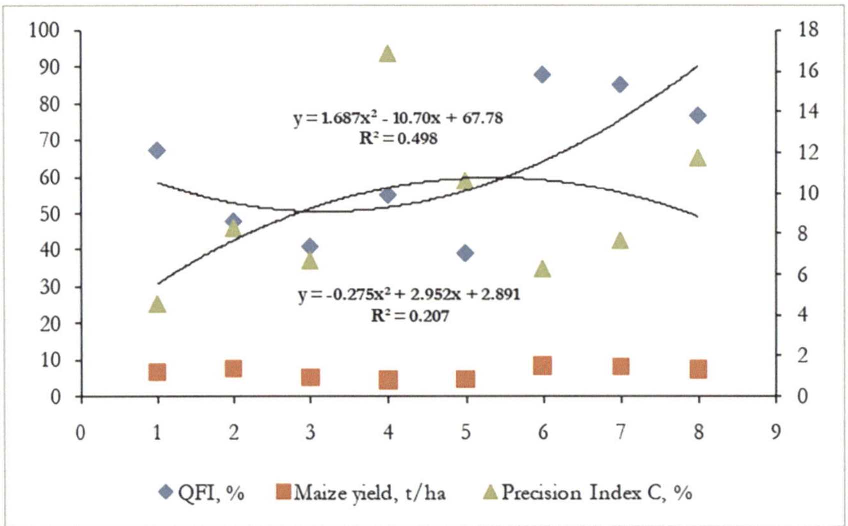

Figure 22.
Maize yield attributed to planter height, parameters like quality of feed index and precision index.

$$y = -0.011x^2 + 0.263x + 6.819 \tag{9}$$

The graphical relation between maize yield and quality of feed index and precision index is shown in **Figure 22**. The prediction equation between quality of feed index (%) and maize yield (t ha^{-1}) was obtained as

$$y = 1.687x^2 – 10.70x + 67.78 \tag{10}$$

The prediction equation between precision index (%) and maize yield (t ha^{-1}) was obtained as

$$y = -0.275x^2 + 2.952x + 2.891 \tag{11}$$

The irrigation water was certainly affected by height of bed and plant population which was related to type of planter used. From all the planters under experiment the pneumatic raised bed (125.15 cm ha^{-1}) and raised bed inclined plate planter (132.75 cm ha^{-1}) recorded minimum water requirement. Thus bed height ranging between 150 and 230 mm (6″–9″) was optimum for irrigation water saving and optimum yield. The highest irrigation water requirement (164.00 cm ha^{-1}) was observed for flood irrigation under flat planting system and lowest yield was recorded for flat planting system (4.98 t ha^{-1}). Raised bed vertical plate planter observed higher irrigation water (148.13 cm ha^{-1}) and lower yield (6.75 t ha^{-1}) than raised bed inclined plate planter practice (132.75 cm ha^{-1}and 7.71 t ha^{-1}). The ridge planting method had water requirement of 145.38 cm ha^{-1} and lower yield. Generally it was found that that more applied irrigation water has inverse relation on maize yield i.e. water at root zone must be not more than sufficient for optimum crop establishment, growth and higher yield. Along with this alternate irrigation ensures more soil aeration and better root growth and underground water saving. Groundwater accounts for 99% of all liquid freshwater on Earth and is present beneath Earth's surface in rock and soil pore spaces and in the fractures of rock formations. Therefore it is very important to make smarter use of the potential of still sparsely developed groundwater resources, and protecting them from pollution and overexploitation and it is essential to meet the fundamental needs of an ever-increasing global population, to address the global climate and energy crises". To achieve the Sustainable Development Goals (SDGs) by 2030 we have to improve the ways for using and managing groundwater efficiently with minimum waste and pollution [76]. Among many contributors to the Polar motion (PM) excitation trend, groundwater storage changes are estimated to be the second largest (4.36 cm/yr) toward 64.16°E [77]. The unregulated anthropogenic activities (like municipal, industrialization, pollution, deforestation, urbanization, building dams, improper landfill practices improper chemical, product, fuel storage causing leaks in soil, agricultural, marine dumping, oil leaks and spills, radioactive waste, global warming killing water animals and thus water pollution, etc.) have drastically increased groundwater depletion and resultant pollution. Groundwater quality monitoring should be done, especially by industries to measure groundwater parameters like Ph, TSS, water level, flow rate, etc. through a telemetry system and if any problem is observed, prompt action should be taken. Climate change will further exacerbate groundwater challenges by affecting aquifers both quantitatively and qualitatively. Geogenic factors such as salinity, fluoride, arsenic and iron in groundwater affect the resource and cause significant long-lasting, intergenerational health detriment. Metals such as cobalt (Co), copper (Cu), iron (Fe), molybdenum (Mo), manganese (Mn) and zinc (Zn) are critical for plant growth and are classified as essential micro nutrients. Other metals that are commonly found as contaminants, and are non-essential for plants, include arsenic (As), cadmium (Cd), chromium (Cr), mercury (Hg), nickel (Ni), lead (Pb), selenium (Se), uranium (U), vanadium (V), Wolfram (W). Metals can have toxic effect on plants even at low concentration. The pollution and depletion of groundwater notoriously violate the right to access water and, in turn, the right to life, recognized as a human right by numerous judicial pronouncements. Water pollution laws must create sufficient legal safeguards against groundwater pollution. The water crises, draught are becoming more common place around the world as billions of people (approx. 6.04 bn) continue to suffer from a lack of access to clean water, sanitation and hygiene in the event of natural water resources scenario in world disasters and increasing global water withdrawals due to growing demand. Projected global water consumption by 2040 is 1.72 trm^3 and highest water consuming sector worldwide by 2050 will be irrigation to agricultural crops. Maize being a C4 plants, has a competitive edge over C3 plants. C4 plants use 3-fold less water, allowing them to grow in conditions of drought, high temperature, and carbon dioxide limitation. Along with this the resource conservation techniques like raised bed planting of crops on raised bed, drip/sprinkler irrigation systems, underground pipeline system (to save evaporation, seepage losses as compared to open channels), crop rotations, rooftop (on building roof) and on farm rainwater harvesting structures (for underground water recharge as well as for use in agricultural lands, industrial, rural and urban area), crop diversification (pulses, sugarcane, maize, etc., in place of rice), agro-forestry, etc. will play a crucial role in preventing over-exploitation of existing water resources and saving of underground water and mitigating climate change. Over-exploitation or pumping groundwater aggressively may release arsenic into the water and also cause land subsidence (sudden sinking of land). Arsenic is mainly present in clayey layer of underground surface and little of it seeps into the water, while groundwater is pumped. But if overdone, a substantial amount may get entered into aquifers due to high hydraulic gradient created. Similarly, phytoremediation

Title	Tractor 45-50HP	Raised bed vertical plate planter	Flat inclined plate planter	Raised bed inclined planter	Pneumatic raised bed/ flat planter	Ridger + manual/ manual planter	Pneumatic flat planter
New cost, P, Rs Cost, USD	550,000 **$6875.00**	60,000 **$750.00**	50,000 **$625.00**	80,000 **$1000.00**	200,000 **$2500**	15,000 **$187.50**	180,000 **$2250.00**
Life (yrs), L	15	10	10	5	10	10	10
Avg. use/yr (h)	700	700	700	300	700	250	700
Rate of interest (%), i	12	12	12	12	12	12	12
Field capacity, ha/h	Of implement	0.49	0.48	0.60	**0.50**	0.56	0.6
Salvage value, S = 10% of P	55,000	6000	5000	8000	20,000	1500	18,000
Total fixed costs (Rs/h)	114.71	15.09	12.57	70.93	50.29	10.56	42.56
Total variable cost (Rs./h)	77.41	446.34	408.62	443.58	297.54	474.75	215.11
Total cost (fixed + variable) (Rs/h)	192.12	461.42	421.19	514.52	347.82	485.31	260.36
Total cost, Rs/ha including tractor		1333.76	1277.74	857.53	695.64	1209.70	433.94
Labour required off machine operation, man-h/ha		10	10	10	10	160	10
Grand total machine cost, Rs/ha Cost, USD*		1646.26 $20.58	1590.24 $19.88	1170.03 $14.62	1008.14 $12.60	6209.70 $77.62	746.44 $9.33

**https://www.xe.com/currencyconverter/convert/?From=USD&To=INR (1USD = 80.00 inr)*

Table 8.
Cost economics calculations for various row crop planters.

(with poplar and other trees, etc.) technique can be used which involves use of plants and associated microbes to reduce the concentrations or toxic effects of contaminants in the environment. However, it is limited to root zone of plant and has limited application where the concentrations of contaminants are toxic to plants. The processes affecting the quality are dissolution, hydrolysis, precipitation, adsorption, ion-exchange, oxidation, reduction and bio-chemical mediated reactions. In general, the reactions that control the chemistry of ground water are:

- Introduction of CO_2 gas into the unsaturated zone.
- Dissolution of calcite and dolomite and precipitation of calcite.
- Cation-exchange.
- Oxidation of pyrite and organic matter.
- Reduction of oxygen, nitrate and sulfate with production of sulphide.
- Reductive production of methane.

Particulars	Raised bed vertical plate planter	Flat inclined plate planter	Raised bed inclined plate planter	Pneumatic raised bed planter	Ridger + manual	Manual	Pneumatic flat planter
	P_{rbvp}	P_{fip}	P_{rbip}	P_{prbvp}	R_{ms}	M_{VP}	P_{pfvp}
Biocides	500.00	500.00	500.00	500.00	500.00	500.00	500.00
Fertilizer	11217.50	11217.50	11217.50	11217.50	11217.50	11217.50	11217.50
Electricity	5737.50	6324.75	5130	4826.25	5602.50	6324.75	6324.75
Seed	1000.00	1000.00	1000.00	1000.00	1000.00	1000.00	1000.00
Human energy	237.52	237.60	236.79	235.15	531.03	373.92	236.79
Machine energy	3443.92	3430.40	3467.80	3400.38	3400.38	3372.67	3430.40
Diesel fuel	39372.48	39325.67	39249.42	39225.96	39256.12	38803.63	38803.63
Total energy MJ ha^{-1}	61508.92	62035.92	60801.51	60405.24	61507.53	61592.47	61513.07
GJ ha^{-1}	61.51	62.04	60.80	60.41	61.51	61.59	61.51
Yield kg ha^{-1}	6750	4980	7710	8610	4970	7420	8340
Specific Energy, MJ kg^{-1}	9.11	12.46	7.89	7.02	12.38	8.30	7.38
Energy productivity, EP, kg MJ^{-1}	0.11	0.08	0.13	0.14	0.08	0.12	0.14

Machine equivalent 133 MJ/kg (Source: CIGR Handbook of Agricultural Engineering Volume V Energy and Biomass Engineering, p. 18).

Table 9.
Energy consumption in maize production.

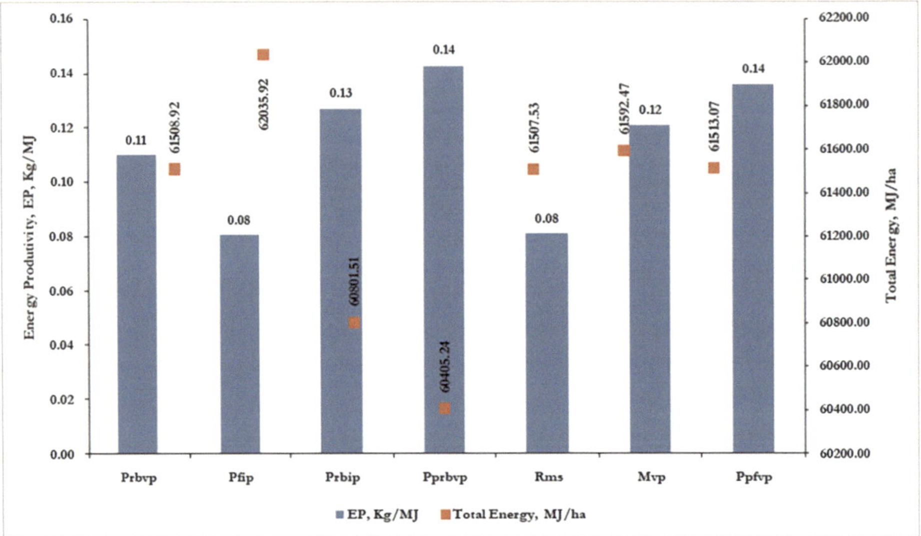

Figure 23.
Total energy and energy productivity associated with various maize sowing planters/techniques.

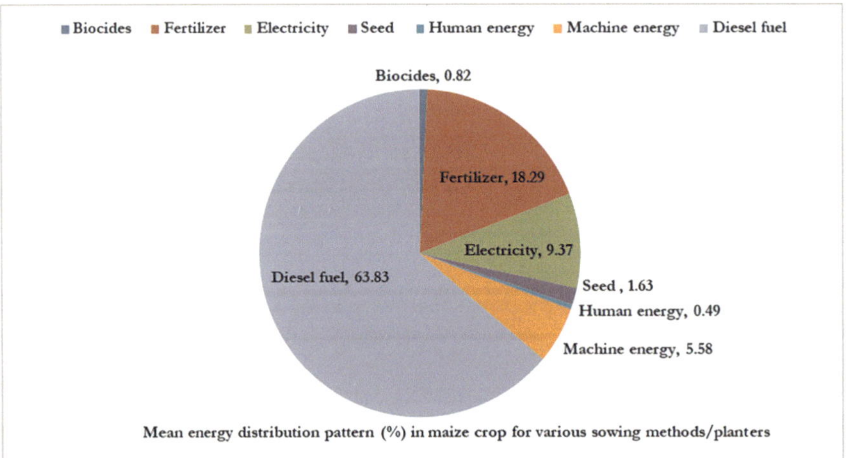

Figure 24.
Maize energy distribution pattern (%) in maize crop for various sowing methods/planters.

- Dissolution of gypsum, anhydrite and halite.
- Incongruent dissolution of primary silicates with formation of clays.

Ground water that is in perpetual motion, acquires various physical, chemical, and biological characteristics as it flows from recharge area to the discharge area. The factors that influence ground water quality are: local geology, land use, climatic conditions particularly pattern and frequency of rainfall and anthropogenic activities such as use of fertilizers and pesticides in agriculture, disposal of domestic sewage and industrial effluents and extent of exploitation of ground water resources.

3.1 Cost economics

The cost economics of the different methods were worked out for pneumatic raised bed planter, vertical plate bed planter, flat inclined plate planter, raised bed inclined plate planter and ridger + manual which are presented in **Table 8**.

The cost of maize sowing operation was found highest for ridger and manual sowing method showing a cost value of Rs. 6209.70 per ha ($ 77.62 per ha) and lowest for pneumatic flat planter showing a cost value of Rs. 746.44 per ha ($ 9.33 per ha) followed by pneumatic raised bed planter as Rs. 1008.14 per ha ($ 12.60 per ha).

The energy calculation for various row crop planters/sowing techniques is shown in **Table 9** and energy productivity is shown in **Figure 23** and pattern is represented in **Figure 24**.

The energy involved was found maximum as 62.04 $GJ.ha^{-1}$ for flat inclined plate planter and energy productivity was lowest for ridger+manual method and flat inclined plate planter as 0.08 kg MJ^{-1} . The specific energy was found minimum for pneumatic raised bed planter as 7.02 MJ kg^{-1} followed by pneumatic flat planter as 7.38 MJ kg^{-1} and raised bed inclined plate planter as 7.89 MJ kg^{-1}. The specific energy for maize sowing was found maximum for flat inclined plate planter as 12.46 MJ kg^{-1}. The energy productivity was found maximum for pneumatic raised bed planter, pneumatic flat planter as 0.14 kg MJ^{-1} followed by raised bed inclined plate planter as 0.13 kg MJ^{-1}.

The major % contribution factor for total energy was diesel fuel (63.83%) in various row crop planters followed by fertilizer (18.29%) and electricity (9.37%). The higher diesel fuel energy is due to more mechanized operations involved in maize cultivation. The machine energy contributed 5.58% in total energy as shown in **Figure 24**.
The variation in electricity energy required for irrigation can be attributed to design of planters and bed shapes variation in various planters. The energy associated with weedicide can be reduced by use of mechanical weeders. Similarly fall armyworm and other insects can be controlled naturally by birds. To give birds a shelter 5–10% of cultivable land should be permanently brought under tree like *Neem* (*Azadirachta indica*), Ashoka tree (*Asopalav*), Tamarind, Jamun tree (*Syzygium cumini*), Banyan (*Ficus benghalensis*), fast growing bamboo (*bambusa vulgaris, Bambusoideae*), Stone apple or aegle marmelos (*bilwa or bael*), Moringa oleifera (drought tolerant), *amla* or Indian gooseberry (*Phyllanthus emblica*), Sal (Shorea robusta), Cedrus deodara, the deodar cedar, Himalayan cedar and Teak (Tectona grandis) tropical hardwood tree species, orchard (mango, guava, apple, kinnow, etc.), etc. Moreover tree act as a carbon capture and storage (CCS) and carbon capture and utilization unit (CCU). Bamboo plants have potential to convert barren lands into a fertile forest. The researchers, from the Mizoram University in Aizawl, India, found that above-ground biomass in the stands of two bamboo species—Bambusa tulda (BT) and Dendrocalamus longispathus (DL)—have

Figure 25.
*Maize crop intercropped with Poplar (*Populus deltoides*) as a mitigation to heavy rainfall, cyclones and floods and also as a diversification option to rice crop in coastal areas.*

Figure 26.
View of agriculture and forest land (Agroforestry) and on-farm rainwater water harvesting and recharging structure (for hilly and flat terrains).

high potential for storing atmospheric carbon. On an average, one hectare of bamboo stands absorbs about 17 tonnes of carbon per year [78]. If planted at optimum distance, tree also helps in natural groundwater recharge. In a study field data from wick lysimeters revealed that the percentage of the yearly rainfall percolating to 1.5m soil depth reached its maximum of 16% of the annual rainfall around the edge of the tree canopy, 4.4m from the nearest tree stem, and decreased to 1.3% in open areas, 37 m away from the nearest tree. The model was run repeatedly and valid for a tree density of 20 trees ha^{-1}, average tree size (67 m^2 canopy area), and 50% water uptake below 1.5 m soil depth [79]. Also during cyclones, storms, trees can protect the properties from debris attack and protect the structures situated downwind from damage . So selection of proper cyclone resistant tree species like *Terminalia arjuna*, *Azadirachta indica*, *Millettia pinnata* (L.) Panigrahi etc. is necessary in coastal areas [80]. Some farmers are practicing maize crop intercropping with poplar (*Populus deltoides*) tree for timber purpose which yields timber around after 5–6 years and good profit to farmers (**Figure 25**). A view of agro forestry concept and rainwater storage and harvesting structure is shown in **Figure 26**. On-farm rainwater harvesting structure can be used at hilly terrains at higher altitude than fields, or in flat terrains if the agricultural land is under organic practices.

The plants are planted at a spacing of 600 cm × 180 cm (20′ × 6′) and with a population of 1000–1250 per ha if grown alone and 750 per ha if grown with some field crops like maize, wheat or turmeric etc. The cost of planting is 25,000 per ha (USD 313 \$ ha^{-1}) and net returns vary between 10.0 and 12.5 lakh per ha (USD 12,500–15,625 \$ ha^{-1}) depending upon growth and girth of plant. Normally this tree grows to height of 85 feet and 36 inches in diameters and average weight of tree ranges between 80 and 120 kg (0.08–0.12 tonne). The average selling price ranges between Rs 12,000–13,000 per tone (USD 150–162 \$ ha^{-1}).

Maize grown in this way can be used for both grain and silage purpose. The populous deltoids tree can tolerate annual precipitation in the range of 600–1500 mm and more making it suitable for flood tolerance [81]. This means that the maize crop intercropping with high water requiring plants like Populus deltoids can be a good mitigation measure in heavy rainfall, flood prone, coastal areas like north-east, north-west and other zones in India and other regions. The water use of a Eucalyptus (2500 l/year) plantation and other tree species such as *Acacia auriculiformis* (1200–1300 litres/year), *Dalbergia sissoo* (1400–1600 litres/year), *Albizzia lebbek* (1200–1300 litres/year) is high. Permanent plantation of such high water demanding tree along with agricultural crops or as plants alone (in 5–10% of cultivable land by every farmer) can help mitigate the climate change effects in flood prone, coastal

Figure 27.
Maize crop is commonly sold on roadside after heat processing and is a good nutrition source.

areas. Farmers can take advantage by selling timber also but plantation area should be maintained by again planting tree on same or new location of cultivable land for combating heavy rainfall, floods etc. Similarly eco friendly technique helps natural control of insects and pests. The eco friendly "Push-pull climate smart" technology entails using an attractive trap plant (Napier/Brachiaria grass as a "pull") and a repellent intercrop (Desmodium as a "push"). Around maize farms, the Napier grass is which attracts stemborers and fall armyworm (FAW) to lay eggs on it but it does not allow larvae to develop on it due to poor nutrition; so very few larvae survive. At the same time, Desmodium, planted as an intercrop emits volatiles that repels stemborers or FAW [82]. Thus energy associated with machine, diesel, electricity and various other inputs can be reduced by selection of appropriate maize planter, climate smart technologies like Push-pull along with Agro-Forestry concept and total energy involved in maize production can also be reduced in a sustainable way and also organic concept can be boosted. Maintaining Agro forestry, birds i.e. biodiversity concept can be useful for other fields crops also. They can protect field crops from excessive heat waves occurring due to climate change and from various insect pests through increased birds population, thus increasing economy of small and marginal farmers. The maize crop can be economical as it creates opportunity from low income families and they buy it from local market and sell them as roadside food on good prices between Rs 20–60 (0.25–0.75 US$) (**Figure 27**).

4. Conclusions

The result reveals that optimum height of bed for better maize crop stand shall range between 150 and 230 mm with a top width of 350 mm bed at a row spacing of 675 mm. The planter plate design geometry has an important role in achieving accurate plant to plant spacing. The yield for inclined and vertical plate mechanism ranged between 4.96–7.71 t ha^{-1} and 6.75–8.61 t ha^{-1} respectively. The saving in water was 9.68–23.69% with bed heights ranging between 150 and 290 mm. The maximum saving in water of 38.85 cm per ha was found for bed height of 150 mm (for 2 rows on as compared to flat planting method. The precision indices for inclined and vertical plate mechanism varied between 4.63–6.74% and 6.35–11.82% respectively. The pneumatic raised bed and flat planter recorded highest yield as 8.61 t ha^{-1} and 8.34 t ha^{-1} respectively. The energy productivity was found maximum for pneumatic raised bed planter, pneumatic flat planter as 0.14 kg MJ^{-1} Maize residue can be collected with balers for use in biomass co generation plants, bio CNG plants, biomass pallet industry as maize crops residue has higher gross calorific value (17.0 MJ kg^{-1}) than paddy crop residue (14.5 MJ kg^{-1}).

Maize crop residue can be used to promote silage industry as farmers usually require silage for feeding animals. The maize crop sowing, weeding and harvesting operations are fully mechanized whereas in case of rice crop manual transplanting is mostly followed in rice growing regions though harvesting is done with combine harvesters. Also the in-situ management of paddy crop residue is energy intensive and maize crop residue can be easily chopped and incorporated with disc harrows, rotary tillers or super seeders facilitating timely and easy sowing of next crops. Among plant-based foods, rice is largest contributor of green house gas emissions, because it can grow in water, so many farmers flood their fields to kill weeds, creating ideal conditions for certain bacteria that emit methane. Rice produces 12 percent of the total greenhouse gas emissions from the food sector, followed by wheat (5%) and sugar cane (2%) [83]. Although burning of straw residues emits large amounts of CO_2, this component of the smoke is not considered as net GHG emissions and only concludes the annual carbon cycle that has started with photosynthesis. At constant straw moisture of 10%, the mass-scaled emission factors (EF_m) were 4.51 g CH_4 and 0.069 g N_2O per kg dry weight ($kg^{-1}dw$) of straw. This corresponds to 1.05% and 0.29% of the total C and N released from straw burning, respectively and subsequent area-scaled emissions (Ea) that were 10.04 kg CH_4 ha^{-1} and 0.154 kg N_2O ha^{-1} as averages for both seasons [60]. Methane in the Earth's atmosphere is a strong greenhouse gas with a global warming potential (GWP) 84 times greater than CO_2 in a 20-year time frame. Methane primarily leaves the atmosphere through oxidization, forming water vapor and carbon dioxide. So, not only does methane contribute to global warming directly but also, indirectly through the release of carbon dioxide. Moreover CH_4 production from rice fields and burning of rice residues also creates breathing problems to local people. The puddled rice also hinders natural recharging of underground water during rainy season (especially *monsoon* period) due to presence of hard pan beneath soil. However strategically diversifying rice area partially to maize crop especially in *Kharif* season can help maintain underground water as well as facilitate recharging also and reducing GHG emissions from its cultivation and residue burning. Maize crop can be sown in *Kharif* (period June-July to October) to diversify rice, *Rabi* season (October to November sowing and harvesting April to June), Spring (sowing-January end to February and harvesting in June-July) and can also be intercropped with Populus deltoids in flood, heavy rainfall prone areas. *Rabi* season or winter maize takes more time to mature as in winter growth of maize is slow but it is less infested with insects, pest, weeds and ensures more efficient use of resources, higher yield than *Kharif* maize and also allows maize-maize system intensification. The rice is grown mainly in *Kharif* season, therefore maize crop can be grown in *Kharif* season to save water. Moreover winter and spring maize have irrigation requirement higher than *Kharif* crop. Also by changing metering plates of pneumatic raised bed planter and inclined plate planter along with some adjustments these planters can be used for sowing of wide row crops like peas, gram, canola etc and narrow row crops like onion, radish, carrot etc. in subsequent winter (*Rabi*) season. With appropriate raised bed maize planter selection, maize sowing operation can be done with precision and lower energy input while maintaining crop yield and saving energy and irrigation water especially for arid and tropical regions. More if agroforestry concept is scaled up, it will help improve water quality, as trees improve water quality by slowing rains as it falls to earth, and helping it soak into the soil. Trees then serve as natural sponge, collecting and filtering rainfall and releasing it slowly into streams and rivers. Trees are the most effective land cover with various benefits such as maintenance of water quality, recharging of water table, reduced drinking water treatment costs, removal of nitrogen and phosphorus leaching from adjacent agricultural land uses into streams by acting as a filtering sediment and also tree can help control the effects of climate change by capturing green house gases and controlling the rise in temperature of earth. Moreover less water requiring crops like pulse, sugarcane, maize, etc. in place of rice will need less irrigation water and more trees can help lower down the environment temperature and more rainfalls. Thus, all these will lead to less pumping of water and saving of underground water as well as natural recharging of underground water.

Acknowledgements

This research was funded and supported by the Department of Farm Machinery and Power Engineering, College of Agricultural Engineering, Punjab Agricultural University, Ludhiana, Punjab, India under the scheme called Establishment of Department of Farm Machinery and Power Engineering, Plan-17-B (PC-1062.2). The team acknowledges School of Climate Change and Agricultural Meteorology, Punjab Agricultural University, Ludhiana, India for providing necessary meteorological data.

Competing interests

This is to certify that all authors do not have any conflict of interest in publishing paper entitled "Scaling mechanization and profitability in maize cultivation through innovative maize planters along with Agroforestry approach -Sustainable and climate smart approach to diversify rice based cereal systems in various regions-" in New Prospects of Maize.

Abbreviations

FIRB	furrow irrigated raised bed
PRB	permanent raised bed
GNSS/IMU	Global Navigation Satellite System
GHG	green house gas
GWP	global warming potential

Author details

Rupinder Chandel*, Mahesh Kumar Narang and Surinder Singh Thakur
Department of Farm Machinery and Power Engineering, Punjab Agricultural University, Ludhiana, Punjab, India

*Address all correspondence to: rupinder26@gmail.com

References

[1] Anonymous. World Maize Scenario – ICAR-Indian Institute of Maize Research. 2022

[2] FAOSTAT. Production-Crops (2020) Data. Food and Agriculture Organization of the United Nations; 2020. Available from: http://www.faostat.fao.org/site/567

[3] Fischer RA, Byerlee D, Edmeades GO. Crop Yields and Global Food Security: Will Yield Increase Continue to Feed the World? Canberra: Australian Centre for International Agricultural Research; 2014. p. 634

[4] Yadav OP, Hossain F, Karjagi CG, Kumar B, Zaidi PH, Jat SL, et al. Genetic improvement of maize in India—Retrospect and prospects. Agricultural Research. 2015;**4**:325-338

[5] Anonymous. Maize vision 2022. A Knowledge Report. 2018. Available from: http://ficci.in/spdocument/22966/India-Maize-Summit.pdf

[6] Kumar R, Srinivas K, Sivaramane N. Assessment of the maize situation, outlook and investment opportunities in India. In: Country Report – Regional Assessment Asia (MAIZE-CRP), National Academy of Agricultural Research Management. Hyderabad, India; 2013

[7] Singh N, Rajendran RA, Shekhar M, Jat SL, Kumar R, Kumar RS. Rabi Maize: Opportunities Challenges. New Delhi: Directorate of Maize Research; 2012. p. 32

[8] Figueiredo M, Cruz I, Silva RBD, Foster JE. Agron. Sustainable Development. 2015;**35**:1175-1118

[9] Porter PM, Hicks DR, Lueschen WE, Ford JH, Warnes DD, Hoverstad TR. Corn response to row width and plant population in the northern Corn Belt. Journal of Production Agriculture. 1997; **10**:293-300

[10] Naresh RK, Singh B, Singh SP, Singh PK, Kumar A, Kumar A. Furrow irrigated raised bed (firb) planting technique for diversification of rice-wheat system for western IGP region. International Journal of Life Sciences Biotechnology and Pharma Research. 2012;**1**(3):134-141

[11] Zhang X, Ma L, Gilliam FS, Wang Q, Li C. Effects of raised-bed planting for enhanced summer maize yield on rhizosphere soil microbial functional groups and enzyme activity in Henan Province, China. Field Crops Research. 2012;**130**:28-37

[12] Ghani A, Greg H, Steven R. Permanent Raised Bed Configurations and Renovation Methods Affect Crop Performance © 2010 19th World Congress of Soil Science, Soil Solutions for a Changing world 1–6 August 2010. Brisbane, Australia; 2010. pp. 171-174

[13] Buehring N. Professor of Agronomy, MSU Conferences/2011-conservation-tillage-conference-proceedings/Corn presentations/ Bed Height And Bed Longevity Effect on Corn and Soybean Yield, 2011; pp. 20–21. Available from: https://www.cottoninc.com/cotton-production/ag-research/agronomy/conservation-tillage/https://www.cottoninc.com/wp-content/uploads/2017/03/Bed-Height-And-Bed-Longevity-Effect.pdf

[14] Naresh K, Rathore RS, Kumar P, Singh SP, Singh A, Shahi UP. Effect of precision land leveling and permanent raised bed planting on soil properties, input use efficiency, productivity and profitability under maize

(*Zea mays*) – wheat (*Triticumaestivum*) cropping system. Indian Journal of Agricultural Sciences. 2014;**84**(6):725-732

[15] Kandil EE, Abdelsalam NR, Mansour MA, Ali HM, Siddiqui MH. Potentials of organic manure and potassium forms on maize (*Zea mays* L.) growth and production. Scientific Reports Nature Research. 2020, 2020:1-11

[16] Kimura T, Bhattacharya KR, Ali SJ. Discoloration characteristics of rice during parboiling. Journal of Society Agricultural Structure. 1993;**24**:23-30

[17] Bhisea SR, Kaura A, Manikantanb MR. Moisture dependent physical properties of maize (*PMH-1*). Acta Alimentaria. 2014;**43**(3):394-401. DOI: 10.1556/aalim.43.2014.3.5

[18] Memon SQ, Mirzam BB, Mari GR. Tillage practices and effect of sowing methods on growth and yield of maize crop. Ağrı. 2007;**40**:89-100

[19] Molin JP, D'Agostini V. Development of a rolling punch planter for stony soil conditions. Agricultural Mechanization in Asia, Africa and Latin America. 1996;**27**(3):17-19

[20] Molin JP, Bashford LL, Bargen VK, Leviticus LI. Design and evaluation of a punch planter for no-till systems. Transactions of ASAE. 1998;**41**(2):307-314

[21] Celik A, Ozturk I, Way TR. Effects of various planters on emergence and seed distribution uniformity of sunflower. Applied Engineering in Agriculture. 2007;**23**(1):57-61

[22] Ani OA, Uzoejinwa BB, Anochili NF. Design, construction and evaluation of a vertical plate maize seed planter for gardens and small holder farmers. Nigerian Journal of technology (Nijotech). 2016;**35**(3):647-655

[23] Yan B, Zhang D, Li Y, Cui T, Yu Y, Tian C, et al. Design and experimental study on small maize planters for Southwest China, ASABE Annual International Meeting. Joseph, Michigan: The American Society of Agricultural and Biological Engineers, St; 2017. p. 1700430. DOI: 10.13031/aim.201700430. Available from: www.asabe.org

[24] Virk SS, Fulton JP, Porter WM, Pate GL. Row-crop planter performance to support variable-rate seeding of maize. Precision Agriculture. 2020;**2020**(21): 603-619

[25] Griepentrog HW. Seed distribution over the area.EurAgEng. 98-A-059, Oslo. International conference on agricultural engineering, Oslo 24-27 August 1998, Norges Landbrukshoegskole, NLH, Aas (Norway); Commission Internationale de Genie Rural, CIGR, Merelbeke (Belgium); European Society for Agricultural Engineers, EurAgEng.- Aas (Norway): Norges Landbrukshoegskole, 1998. v. 1: 530 p.; v. 2: p. 533-1064

[26] Panning JW, Kocher MF, Smith JA, Kachman SD. Laboratory and field testing of seed spacing uniformity for sugarbeet planters. Applied Engineering in Agriculture. 2000;**16**(1):7-13

[27] Jasa PJ, Dickey EC. Tillage factors affecting corn seed spacing. Transactions of ASAE. 1982;**25**(6):1516-1519

[28] Campbell CM. Influence of seed formation of corn on accumulation of vegetative dry matter and stalk strength. Crop Science. 1964;**4**(1):31-34

[29] Duncan WG. Maize. In: Evans LT, editor. Crop Physiology: Some case Histories. Cambridge University Press;

1975. pp. 23-50. Available from: http://books.google.com/books?id=Yro8AAAAIAAJ&lpg=PR6&ots=FRaZ9zN_SG&dq=crop%20physiology%20some%20case%20histories%20duncan&lr&pg=PA36#v=onepage&q&f=false

[30] Widdicombe WD, Thelen KD. Row width and plant density effects on corn grain production in the northern corn belt. Agronomy Journal. 2002;**94**(5):1020-1023

[31] Maddonni GA, Otegui ME, Andrieu B, Chelle M, Casal JJ. Maize leaves turn away from neighbors. Plant Physiology. 2002;**130**(3):1181-1189

[32] Robinson RG, Ford JH, Lueschen WE, Rabas DL, Warnes DD, Wiersma JV. Response of sunflower to uniformity of plant spacing. Agronomy Journal. 1982;**74**(2):363-365

[33] Lyu H, Li B, Yu T, Zhao Z, Huang J, Zheng Y. Detection and Control of Intelligent Maize Planter, ASABE Annual International Meeting. St. Joseph, Michigan: The American Society of Agricultural and Biological Engineers; 2017. p. 1700891. DOI: 10.13031/aim.201700891. Available from: www.asabe.org

[34] Erbach DC, Cruse RM, Crosbie TM, Timmons DR, Kaspar TC, Potter KN. Maize response to tillage-induced soil conditions. Transactions of ASAE. 2013;**29**(3):0690-0695. DOI: 10.13031/2013.30214

[35] Yakle GA, Cruse RM. Corn plant residue age and placement effects upon early corn growth. Canadian Journal of Plant Science. 1983;**63**:871-877

[36] Zhang X, He J, Li H, Wang Q, Wu S. Design and experiment on no-till planter in horizontal residue-throwing finger-wheel type for maize. Transactions of the Chinese Society of Agricultural Machinery. 2010;**41**(12):39-43. DOI: 10.3969/j.issn.1000-1298.2010.12.008

[37] Chen H, Huang H, Li H. Design of row-followed no-till wheat and maize planter under controlled traffic farming system. Transactions of the Chinese Society for Agricultural Machinery. 2009;**40**(3):72-76

[38] Gao N, Zhang D, Yang L. Mathematical modeling and Simulation of an Anti-blocking Mechanism for No-till Maize Planter. In: American Society of Agricultural and Biological Engineers Annual International Meeting. St. Joseph, Michigan: American Society of Agricultural and Biological Engineers; 2013. Available from: www.asabe.org

[39] Griepentrog HW, Nørremark M, Nielsen H, Blackmore BS. Seed Mapping of Sugar Beet. Precision Agriculture Journal. 2005;**6**:2157-2165

[40] Smith P, Bustamante M, Ahammad H, Clark H, Dong H, Elsiddig EA, et al. Agriculture, Forestry and Other Land Use (AFOLU). In: Edenhofer O, Pichs-Madruga R, Sokona Y, Farahani E, Kadner S, Seyboth K, et al., editors. Climate Change 2014: Mitigation of Climate Change. Contribution of Working Group III to the Fifth Assessment Report of the Intergovernmental Panel on Climate Change. Cambridge; New York, NY: Cambridge University Press; 2014

[41] IPCC. Technical summary. In: Stocker TF, Qin D, Plattner GK, Alexander LV, Allen SK, Bindoff NL, Bréon FM, Church JA, Cubasch U, Emori S, Forster P, editors. Climate Change 2013: The Physical Science Basis. Contribution of Working Group I to the Fifth Assessment Report of the Intergovernmental Panel on Climate

Change. Cambridge University Press; 2013

[42] Linquist B, Groenigen KJV, Adviento-Borbe MA, Pittelkow C, Kessel CV. An agronomic assessment of greenhouse gas emissions from major cereal crops. Global Change Biology. 2011;**18**(1):194-209

[43] Bhatia A, Aggarwal PK, Jain N, Pathak H. Greenhouse gas emission from rice- and wheat-growing areas in India: Spatial analysis and upscaling. Greenhouse Gases: Science and Technology. 2012;**2**(2):115-125

[44] Qiu J. China cuts methane emissions from rice fields. Nature. 2009

[45] Zhang B, Tian H, Ren W, Tao B, Lu C, Yang J, et al. Methane emissions from global rice fields: Magnitude, spatiotemporal patterns, and environmental controls. Global Biogeochemical Cycles. 2016;**30**(9): 1246-1263

[46] Gupta K, Kumar R, Baruah KK, Hazarika S, Karmakar S, Bordoloi N. Greenhouse gas emission from rice fields: A review from Indian context. Environmental Science and Pollution Research. 2021;**28**:30551-30572

[47] Gadde B, Bonnet S, Menke C, Garivait S. Air pollutant emissions from rice straw open field burning in India, Thailand and the Philippines. Environmental Pollution. 2009;**157**(5): 1554-1558

[48] Gupta P, Sahai S, Singh N, Dixit C, Singh D, Sharma C, et al. Residue burning in rice-wheat cropping system: Causes and implications. Current Science. 2004;**87**(12):1713-1717

[49] Miura Y, Kanno T. Emissions of trace gases (CO_2, CO,CH_4, and N_2O) resulting from rice straw burning. Soil Science Plant Nutrition. 1997;**43**(4): 849-854

[50] Anonymous. Climate change indicators: Greenhouse gases | US EPA. 2021

[51] World Bank. South Asia's Hotspots: Impacts of Temperature and Precipitation Changes on Living Standards. Washington, DC: World Bank; 2018

[52] Naser HM, Nagata O, Sultana S, Hatano R. Impact of management practices on methane emissions from paddy grown on mineral soil over Peat in Central Hokkaido, Japan. Atmosphere. 2018;**9**(212):1-18

[53] Smartt AD, Brye KR, Rogers CW, Norman RJ, Gbur EE, Roberts HJT, et al. Previous crop and cultivar effects on methane emissions from drill-seeded, delayed-flood rice grown on a clay soil. Hindawi Publishing Corporation. Applied and Environmental Soil Science. 2016;**9542361**:1-13

[54] Tang H, Xiao X, Tang W, Wang K, Sun J, Li W, et al. Effects of winter covering crop residue incorporation on CH_4 and N_2O emission from double-cropped paddy fields in southern China. Environmental Science and Pollution Research International. 2015;**22**(16): 12689-12698

[55] Lu X. A meta-analysis of the effects of crop residue return on crop yields and water use efficiency. PLoS One. 2020; **15**(4):1-18

[56] Kushwah SK, Dotaniya ML, Upadhyay AK, Rajendiran S, Coumar MV, Kundu S, et al. Assessing carbon and nitrogen partition in Kharif crops for their carbon sequestration potential. Proceedings of the National Academy Science Letter. 2014;**37**(3):213-217

[57] Zomer RJ, Bossio DA, Sommer R, Verchot LV. Global sequestration potential of increased organic carbon in cropland soils. Scientific Reports. 2017;7: 15554. DOI: 10.1038/s41598-017-15794-8.:1-8

[58] Zwart SJ, Bastiaanssen WGM. Review of measured crop water productivity values for irrigated wheat, rice, cotton and maize. Agricultural Water Management. 2004;**69**:115-133

[59] Anonymous. Corn fields help clean up and protect the environment – MSU Extension. 2007. Available from: https://www.canr.msu.edu/news/corn_fields_help_clean_up_and_protect_the_environment

[60] Romasanta RR, Sander BO, Gaihre YK, Alberto Ma C, Gunmert M, Quilty J, et al. How does burning of rice straw affect CH_4 and N_2O emissions? A comparative experiment of different on-field straw management practices. Agriculture, Ecosystems & Environment. 2017;**239**:143-153

[61] Pandey MM, Sirohi NPS, Singh KK, Ganesan S, Dhingra D, Singh SK, et al. Handbook of Agricultural Engineering. New Delhi: Indian Council of Agricultural Research; 2013. p. 7

[62] Yilmaz I, Akcaoz H, Ozkan B. An analysis of energy use and input costs for cotton production in Turkey. Renewable Energy. 2005;**2005**(30):145-155

[63] Ozkan B, Kurklu A, Akcaoz H. An input-output energy analysis in greenhouse vegetable production: A case study for Antalya region of Turkey. Biomass and Bioenergy. 2004;**26**:189-195

[64] Barber AA. Case Study of Total Energy and Carbon Indicators for New Zealand Arable and Outdoor Vegetable Production. New Zealand: Agricultural Engineering Consultant Agril INK; 2003

[65] Kachman SD, Smith JA. Alternative measures of accuracy in plant spacing for planters using single seed metering. American Society on Agricultural Engineering. 1995;**38**:379-387

[66] Doerge T, Hall T. Wisconsin Fertilizer, Aglime and Pest Management Conference. Madison, WI; 2002

[67] Carlson G, Doerge T, Clay D. Estimating Corn Yield Loss from Unevenly Spaced Plants. International Plant Nutrition Institute Publication; 2003

[68] Liu W, Tollenaar M, Stewart G, Deen W. Impact of planter type, planting speed, and tillage on stand uniformity and yield of corn. Agronomy Journal. 2004;**96**(6):1668-1672

[69] Van Roekel RJ, Coulter JA. Agronomic responses of corn hybrids to row width and plant density. Agronomy Journal. 2012;**104**(3):612-620

[70] Shubeck FE, Young HG. Equidistant plant spacing. Crops and Soils Magazine. 1970;**22**:12-14

[71] Karayel D, Ozmerzi A. Effect of tillage methods on sowing uniformity of maize. Canadian Biosystems Engineering. 2002;**44**:2.23-2.26

[72] Vamerali T, Bertocco M, Sartori L. Effects of a new wide-sweep opener for no-till planter on seed zone properties and root establishment in maize (*Zea mays*, L.): A comparison with double-disk opener. Soil and Tillage Research. 2006;**89**(2):196-209

[73] Altuntas E, Ozgoz E, Taser OF, Tekelioglu O. Assessment of different

type furrow openers using a full automatic planter. Asian Journal of Plant Sciences. 2006;**5**(3):537-542

[74] Kepner RA, Bainer R, Barger EL. Principles of Farm Machinery. Westport, CT: Avi Pub. Co; 1978. p. 221

[75] Hossain I, Hossain I, Mamun MAA, Siddiquie NA, Rahman MM, Sq RM. Two wheel tractor operated strip tillage seeding equipment for dry land farming. International Journal of Energy Machinery. 2012;**5**(1):35-41

[76] Anonymous. 2022. Available form: https://www.unwater.org/un-world-water-development-report-2022/

[77] Seo K-W, Ryu D, Eom J, Jeon T, Kim J-S, Youm K, et al. Drift of Earth's pole confirms groundwater depletion as a significant contributor to global sea level rise 1993–2010. Geophysical Research Letters. 2023;**50**: e2023GL103509. DOI: 10.1029/2023GL103509.PP:1-7

[78] Anonymous. Bamboo plants can act as efficient carbon sinks (nature.com). 2021

[79] Ilstedt U, Tobella AB, Bazié HR, Bayala J, Verbeeten E, Nyberg G, et al. Intermediate tree cover can maximize groundwater recharge in the seasonally dry tropics. Scientific Reports. 2016;**6**:1-12

[80] Arzoo A, Pradhan S. A review on cyclone resistant plants found in cyclone Prone Odisha, India. International Journal of Scientific & Technology. 2020;**9**(4):488-491

[81] Rodriguez ME, Lauff D, Cortizo S, Luquez VMC. Variability in flodding tolerance, growth and leaf traits in a *Populus deltoids* intraspecific progeny. Tree Physiology. 2019;**40**:19-29

[82] Anonymous. Integrated management of the Fall Armyworm on maize A guide for Farmer Field Schools in Africa. Food and Agriculture Organization of the United Nations Rome, 2018. 2018. Available from: https://www.fao.org/documents/card/en/c/I8741EN/

[83] Anonymous. 2021. Available from: https://www.greenbiz.com/article/how-much-do-crops-contribute-emissions

Chapter 6

Agricultural Transformation in Maize Producing Areas of Africa

Paul L. Woomer, Dries Roobroeck and Welissa Mulei

Abstract

Maize is a critical staple cereal across Sub-Saharan Africa but attempts to improve its productivity in small-scale farming systems often prove disappointing. The 12 key technologies required to overcome poor yields are mostly known, but the manner in which they are mobilized, packaged, and delivered requires re-evaluation. Combinations of better varieties and their necessary accompanying inputs must become more available and affordable for an African maize revolution to succeed, and land must be managed in ways that enhance, rather than diminish, land quality over time. Adjustments to the bundling and transfer of these technologies as transferable assets pose a solvable dilemma. These interventions must be based upon specific agro-ecological and socio-economic contexts and offered within the scope of well-designed regional and national agricultural development agendas. Success in boosting maize yields and its companion field legumes form the basis for greater food security across Africa and value-adding enterprises, including the growth of blended flours and commercial animal production. This chapter describes how these technologies may be most effectively mobilized within the current thrust to transform African agriculture.

Keywords: agricultural transformation agenda, Dakar 2 feed Africa summit, International Institute of Tropical Agriculture, maize-based cropping systems, rural development, small-scale farming systems, TAAT program, technology packages

1. Introduction

Maize first arrived from the New World in Africa during the 1500s. Portuguese traders supplying fortresses and coastal trading centers first introduced it. Still, the crop quickly appealed to African farmers due to its high yield, low labor requirements, and short growing season. Cultivation swiftly spread because of its higher yield than existing indigenous staples, mostly millet and sorghum, and its ability to function as a substitutable dietary staple [1]. Miracle [2] concluded that the timing when maize became important in different parts of Africa is better understood than the exact introduction sites and parties responsible. In general, maize was first introduced to West Africa, then spread inland and southward, and last reached East Africa and deeper Central Africa.

Maize offered the right combination of traits for widespread adoption. Its nutrients are concentrated and easily transported, husks protect against pests, birds, and extreme weather, and it is consumable over an extended period starting with young cobs through harvest maturity and longer given proper storage and milling [3]. The appearance of maize and its cultivation practices are like sorghum and millet, further encouraging adoption. Many New World crops were introduced to Africa following the discovery of the New World, including cassava, sweet potato, Irish potato, groundnut, and beans. However, maize had the most rapid and greatest impact on native farming systems, and this impact continues into the present [4].

Maize is now Africa's most important crop. About 30% of energy intake across Africa comes from maize [4]. Erenstein et al. [5] provide a recent description of maize production, consumption, and trade, including its trends in different regions and countries of Africa. Maize is the second most cultivated crop worldwide after wheat. It is grown on about 197 million ha, which supplies 1137 million tons of dry grain. Africa accounts for 21% of that land use cover but only 7.4% of production, signifying strong shortfalls in yield. Average maize grain yields in Africa are only 2.1 t ha^{-1} compared to 5.8 t ha^{-1} worldwide and 10.8 t ha^{-1} in North America, and closing this yield gap is critical to realize African food security and nutrition. Various agronomic and economic factors cause attainable yields not to be met by smallholder farmers, ranging from seed quality, low and variable returns on fertilizer investment, nutrient and water availability, pest control, labor, and equipment assets to gender-specific challenges, value addition, market linkage and selling prices [6, 7]. The role of technology access and adoption on production levels is evident at the level of individual farmer fields and entire regions. For instance, plots closest to the homestead typically receive most inputs and are better weeded than out-fields and therefore record higher yields. East and Southern Africa have a smaller maize growing area than West and Central Africa, yet production is higher, which is primarily ascribed to better varieties, fertilizer supply, and extension support.

To safeguard the production and self-sufficiency of this staple, African governments have adopted policies like subsidizing production inputs and various price protection schemes, some benefiting producers and others designed to assure consumers. Some of these measures have been criticized by development banks as unsustainable, leading to structural adjustments, particularly those that resulted in unrealistic consumer prices and fluctuating supply [4]. Some research suggests that Africa should prioritize policies and programs centered on non-monetary incentives like advancing technology and infrastructure, investing in irrigation, precision agriculture, research services, and human development [8–10]. However, a growing consensus is that maize responds favorably to production inputs, particularly hybrid seeds and fertilizers. Programs encouraging investment into making these materials more available to small-scale producers are a promising pathway to rural development in maize-producing areas [11, 12]. This chapter focuses on the needed maize technologies and how they may be more effectively deployed and delivered through emergent developmental strategies to ensure maize's future in Africa.

The African agricultural development community recently consolidated around the Dakar 2 Feed Africa Summit organized by the African Development Bank and held on 22–27 January 2023. Its purpose was to unlock Africa's agricultural potential by delivering climate-smart agricultural technologies to millions of farmers and creating an enabling environment for market-driven economic development through improved value addition, rural infrastructure, and stronger policy incentives. Thirty-four (34) African Heads of State, 75 Ministers, and numerous heads of development organizations attended the Summit. They presented and discussed Country Food and Agriculture Delivery Compacts to further the Feed Africa Strategy [11] at national levels based on revised production targets for key agricultural commodities, planned improvements in enabling policies and rural infrastructure, and options for innovative financing. Following the Summit, the 36th African Union Assembly endorsed its outcomes and called for time-bound and measurable indicators for success. Within a month, the Summit mobilized more than $70 billion in investment to boost food and agriculture production across the continent. IITA's Partnership for Delivery (P4D) staff and Regional Directors are working with country planners to realize the vision of Dakar 2 African agricultural transformation. Follow-up to this event is critical for the timely modernization of agricultural technologies and services employed across Africa, including those related to maize-based cropping systems.

2. Modernizing maize technologies

There are 12 key technologies to modernize production and post-harvest management of maize in Africa. These technologies include: (1) drought-tolerant maize varieties to strengthen the resilience of food production, (2) imazapyr-resistant maize varieties that withstand parasitic striga weeds, (3) golden maize varieties with provitamin A biofortification for improved human nutrition, (4) a streamlined licensing mechanism for commercial multiplication of hybrid maize varieties, (5) information and communication technology (ICT) platforms offering ready access to digital information, (6) contracted farm mechanization services, (7) better fertilizer blends together with top-dressed nitrogen for improved nutrient supply, (8) rotating and intercropping with nitrogen-fixing grain legumes to improve soil health, (9) applying herbicides for pre- and post-emergent weed control, (10) controlling the biological invasion of fall armyworm (FAW) through integrated practices, and (11) countering aflatoxin contamination through atoxic competitors, (12) improved post-harvest handling. Further details on each of these 12 technologies follow.

2.1 Drought-tolerant maize varieties

Recently released maize varieties allow acceptable grain yields under short-term and moderate drought (**Figure 1**). This technology mitigates adverse climate and lessens the risk of crop failure across many zones of Sub-Saharan Africa [13, 14]. Insufficient rainfall is a widespread reason for lost maize yield across Sub-Saharan

Figure 1.
Drought tolerant maize variety (left) and drought sensitive variety (right).

Africa, as 90% of the land is rainfed rather than irrigated. As a result, maize yields are highly sensitive to seasonal rainfall [15]. Improved lines offered through the market include drought tolerant maize (DTMA) with an ability to withstand periods of acute soil drying and water efficient maize (WEMA) adapted to season-long reduced supply of soil moisture.

Timely access to weather and market information and local climate adaptation measures provides a means for sound decision-making regarding when and where to invest in drought-tolerant maize. These same technologies allow maize to be produced in semi-arid regions with less irrigation water, allowing growers and national development planners to utilize less-traditional growing areas better. More than 200 lines of DTMA have been released in 13 African countries, and over 120 hybrids of WEMA released in seven countries. DTMA includes hybrid varieties that require parent seed and licensing [16] and numerous open-pollinated varieties (OPVs). These latter varieties permit royalty-free purchase and multiplication through farmers' and community-based seed production.

These drought-tolerant varieties are introduced to farmers through on-farm demonstrations with broader coverage. This allows farmers to see the varieties in action and learn about their benefits firsthand. The main barriers to adopting DroughtTEGO® varieties are a lack of information about their productivity, unavailability of seed when needed, and the high cost compared to other locally available varieties [17, 18]. Oniang'o et al. [18] found that well-thought-out strategies to influence awareness and adoption of drought-tolerant maize include strengthening extension services, providing credit to small-scale farmers, investigating cases of discontinued use, improving access to seed through agro-dealerships, targeting age and gender, and specific agro-ecological zones.

2.2 Imazapyr-resistant maize for Striga management

Parasitic Striga invades the roots of maize and other cereals to remove water and nutrients from host plants. Maize cannot resist striga (**Figure 2**), causing stunting,

Figure 2.
Severe striga infestation of maize in western Kenya.

abnormal growth, and small ears, resulting in a yield reduction of 30 to 80% [19]. About 20 million ha of farmland in sub-Saharan Africa is Striga-infested, resulting in over US $1 billion per year in yield loss and threatening the food security and livelihoods of over 100 million people [20]. Improved maize varieties that resist the herbicide imazapyr (the IR trait) are becoming available, protecting the roots against parasitic invasion [21]. Very low levels of imazapyr (e.g., 30 to 45 g per ha) are applied to maize seeds. However, the application rate is critical because too much imazapyr can harm maize germination and early growth. When used correctly, the herbicide is placed exactly where and when needed to control striga as it starts invading young maize roots.

Imazapyr herbicides are made from the active ingredient imidazoline, mixed with salt to form a stable powder. The herbicide is then coated onto maize seeds using an adhesive. An example of such a seed treatment system is patented under the term StrigAway. IR maize seeds are planted following recommended soil and fertilizer management practices for a growing area. Imazapyr is non-toxic to mammals, but it is important to wear gloves or wash hands when planting the seed manually, as they may be mixed with insecticides. The spread of this technology across Africa has been slower than expected, given the scope of the problem. Where available, agro-input suppliers sell IR maize seed at about US $3 per kilogram. Yield increases of 1.0 to 3.0 tons of grain per hectare are achieved compared to comparable varieties not protected by imazapyr [22]. An additional benefit is that the Striga seed bank diminishes over time, eventually eradicating striga from croplands [23].

However, this technology is often too expensive for subsistence farmers. A more affordable option is to integrate imazapyr herbicide technology with other measures, such as planting soybeans followed by striga-resistant maize varieties using nitrogen fertilizer and good agricultural practices [24, 25]. The adoption of this technology depends on several factors, including the age and education of the household head, the availability of training and support for farmers, membership in farmer group, the availability of the technology, perceptions based on the social and cultural context, and the political climate [26].

2.3 Vitamin a biofortified maize

Biofortified maize varieties higher in Vitamin A are also becoming more widely available. Maize is a staple food for over 300 million people across Sub-Saharan Africa; however, the widely grown starchy white varieties contain sub-optimal minerals and vitamins. Conventional breeding has improved the content of provitamin A in maize, offering a viable avenue to improve community nutrition. Golden maize contains beta-carotene, lending it a bright orange color (**Figure 3**). These compounds are converted into vitamin A after ingestion. More than 40 of these biofortified varieties have been released across Sub-Saharan Africa [27]. These varieties were originally developed from Central and South American lines naturally rich in provitamin A and then crossed with well-adapted lines holding improved agronomic traits such as disease resistance and drought tolerance.

Unlike biofortified lines, pro-vitamin A is often oxidized and forms off-flavors other maize varieties. Pro-vitamin A biofortified maize offers a cost-effective solution to Vitamin A deficiency in areas where people consume fresh and dried maize [28]. It provides half the daily Vitamin A requirement for adults and costs $0.8 to $1.2 per kg of OPV seed [29]. Golden maize contains 8 to 15 parts per million of pro-vitamin A, while conventional varieties do not have this nutrient. Biofortification of maize is a

Figure 3.
Biofortified maize (center) rich in vitamin a compared to conventional yellow and white varieties (top and bottom).

promising approach to combat micronutrient deficiencies in sub-Saharan Africa. Provitamin A biofortified maize is a safe and effective way to improve vitamin A status, and it has been well-accepted by most communities compared to yellow maize, which has associated negative perceptions. With proper policy support, biofortified maize can help address the deficiency of vitamin A in this region. A study conducted by Nesamvuni et al. revealed that introducing a vitamin-fortified maize meal to the meals of African children aged one to three led to positive effects, such as better weight gain and improvements in specific aspects of their vitamin A levels [30]. Different varieties are available for cultivation in lowland and highland elevations and semi-arid and humid climates.

2.4 Information communication technology (ICT) platforms

Agricultural information is vast and covers many areas of expertise, depending on the specific agro-climatic zones and socio-economic contexts. To ensure its effective use, it is crucial to have a well-organized system for sharing this information. It is equally essential to disseminate the right information to the right people at the right time. Fortunately, with information and communication technology (ICT) advancements, we can leverage these tools to provide farmers with accurate, timely, and relevant information and services. This, in turn, helps them adopt new technologies more effectively and makes their agricultural endeavors more profitable [31]. Based on an analysis conducted by Ayim et al. [32], the primary ICT technologies used in Africa to improve agriculture productivity are text and voice-based services designed for mobile phones. The rise of smartphone technology, including apps, has also led to the development of innovations in the farming industry. Radio and television are also popular tools for sharing agricultural information with rural farmers. These can be as robust and interactive as virtual workshops and webinars where call-in segments or SMS/text interactions are included to address farmer queries in real-time during the broadcasting. Computers are a gadget primarily utilized by researchers. ICT has enabled the development of dedicated websites, social media, mobile-based extension services, financial inclusion and mobile payments, and online communities to share information on agricultural technologies, including available technologies, coverage, best practices, pest and disease management, market prices, success stories, capacity-building events, and implementation guides, while also incorporating interactive features like search functions, discussion forums and chatbot for farmer engagement and query resolution by experts. Such channels include the TAAT website, https://taat-africa.org/, and the TAAT mobile app available on the App Store. Although there are many benefits to using ICT tools and systems in agriculture, most agricultural and farming communities in Africa need to adopt them to the extent necessary for significant agricultural development. Specific factors that hinder the widespread adoption and diffusion of these services include inadequate technological infrastructure, language barrier, affordability, unsuitable ICT policies, lack of awareness regarding the potential contribution of ICTs to the farm business, and a low level of user skills, particularly among farmers [33].

One such digital application relevant to maize-based systems is The Product Platform for Agricultural Solutions (ProPAS), which offers open access information about innovative technologies in English or French. Each profile covers various aspects relating to the problems addressed, functional principles, geographic suitability, composition, application, customer segmentation, capital/operational costs, expected benefits, and licensing (see **Figure 4**). The platform has two goals; to provide technology holders with a means to disseminate their proven and promising solutions and to encourage users to search through options that can assist their agricultural objectives (visit https://propas.iita.org/). The database allows filtering solutions based on multiple search fields such as relevant value chain, its form and type (i.e., genetics, input supply, management, equipment, and digital tools), location where available, and target beneficiaries. Fourteen of its solutions relate to maize, and another fourteen describe leguminous companion crops. In 2022, ProPAS received 27,207 visitors, of which 9.1% were profile views for maize technologies, and overall it attracted the most attention to equipment followed by genetics, management, and input supply.

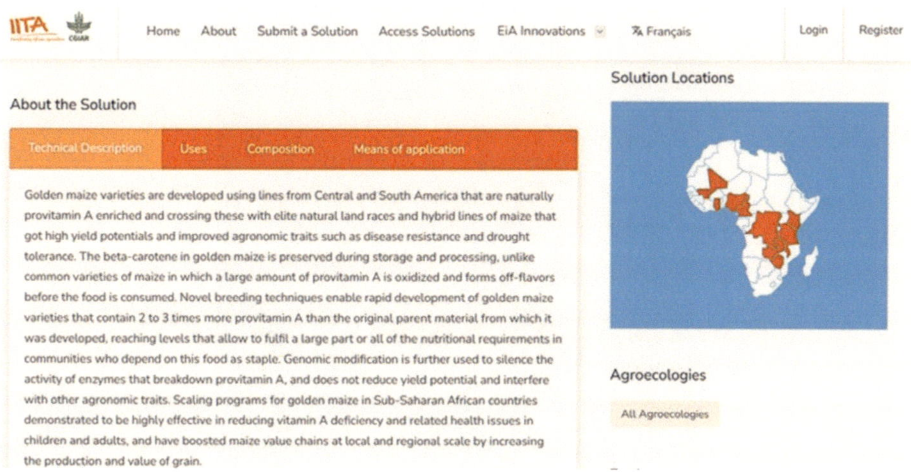

Figure 4.
Output from the ProPAS website describing "golden maize".

2.5 Commercial licensing systems

Limited investment by the commercial seed production sector impedes the availability of improved maize varieties to small-scale farmers across Sub-Saharan Africa. In response, The African Agricultural Technology Foundation (AATF) established a series of public-private ventures for the multiplication of high-yielding, drought-tolerant TEGO® (conventional) and insect-protected TELA® (transgenic) maize hybrids. Seven African countries now produce seeds of these elite varieties, accompanied by a licensing model and agri-business training that now supplies millions of farmers through this mechanism [34]. Precautions are in place to ensure that this multiplication process ensures true-to-type seed with a high germination rate.

Hybrid maize varieties have a high market value and provide opportunities for businesses to generate investment returns from seed multiplication and developing new, improved lines. Significant increases in food and nutritional security and farm incomes are realized where TEGO® (**Figure 5**) [35] and TELA® seed systems are adopted because these varieties produce higher grain yield and quality than other cultivated lines under normal and lower rainfall. Royalty-free licensing results in new, improved maize varieties from public institutions becoming more rapidly available to farmers through commercial transfer rights to the private sector. This mechanism includes an agreement between the holder of intellectual properties for maize varieties and a legally eligible enterprise that intends to multiply and sell these seeds commercially.

Between 2013 and 2020, 7032 tons of Drought TEGO® and 161 tons of TELA® hybrid seeds were sold and planted on an estimated 287,720 hectares of cropland to produce over 1 million tons of grain. This maize is valued at US $236 million, benefiting about 4.3 million people [34]. At the end of 2020, variety licenses were signed with 38 seed companies from seven countries to commercialize these elite TEGO® and TELA® maize hybrids and test new lines. In this way, the TEGO® and TELA® mechanism is intended to streamline the licensing process for elite, climate-smart maize and to link intellectual goods to commercial opportunities.

Figure 5.
TEGO® maize produced under commercial license.

Stress-resilient crop varieties are often seen as inferior to regular hybrids, but this is a misconception. A multi-location evaluation study of stress-resilient maize hybrids by the International Maize and Wheat Improvement Center (CIMMYT) in Sub-Saharan Africa [36] found that stress-resilient maize hybrids produced yields that were on par with or even superior to regular hybrids under both favorable and unfavorable conditions. These hybrids are a good choice for farmers as they help to achieve more stable yields over time and build a more resilient food system.

2.6 Contract mechanization services and applications

An increasing amount and variety of mechanized agricultural services are offered to farmers across Sub-Saharan Africa. Unfortunately, this contracted and rented use of mechanization services remains limited because contracting businesses experience difficulties in informing and convincing lower-income farmers of their value. Ironically, these contracted services provide labor-reducing operations through equipment beyond small-scale farmers' purchasing power [37]. African countries must develop favorable arrangements to make agricultural mechanization accessible to small and medium-scale farmers. This could be done by incentivizing the private sector to scale up agricultural mechanization initiatives and targeting and engaging women farmers and youth by investing in supportive infrastructure and training [38]. ICT applications can help farmers access contract mechanization services by matching farmers with mechanization service providers, providing information about mechanization services, and tracking the progress of mechanization activities. For instance, data on the extent of land cultivated or harvested can be utilized to verify the fulfillment of mechanization services and ascertain whether farmers receive the expected benefits for their investments.

Nevertheless, there is a lot of debate about the role of mechanization and digitalization in African agricultural transformation. A study by Daum et al. [39] documented these concerns by national stakeholders in several African countries. Some argue that mechanization is essential for reducing drudgery, increasing productivity, and reducing poverty, while others say that it can lead to the displacement of rural labor and environmental degradation. Furthermore, there is a continued appeal for state-led mechanization in some countries, even though this approach has been criticized for being inefficient and corrupt. This has resulted in yet another debate about how governments should best promote mechanization in Africa. Some people believe that governments should provide subsidized tractors and run public hire centers, while others believe that the state should focus on creating an enabling environment for private actors.

On the other hand, digitalization is seen as a promising tool, but there are concerns about data sovereignty and the digital divide. Moreover, gender and age can influence how people view digitalization, with younger people and women being more likely to be optimistic about its potential. Therefore, policymakers and development institutions must consider local stakeholders' viewpoints to aid in selecting and designing the most promising policies/programs and ensure their effective implementation at the grassroots level.

Hello Tractor (**Figure 6**) is a success in this area, an award-winning equipment-sharing application that connects tractor operators to African smallholder farmers. This digital platform results in the collaborative use of mechanized field operations by creating a common marketplace between machine owners and farmers who request and pay for services via messaging. The smartphone application also supports credit scoring and provides market intelligence for risk management and loan repayment.

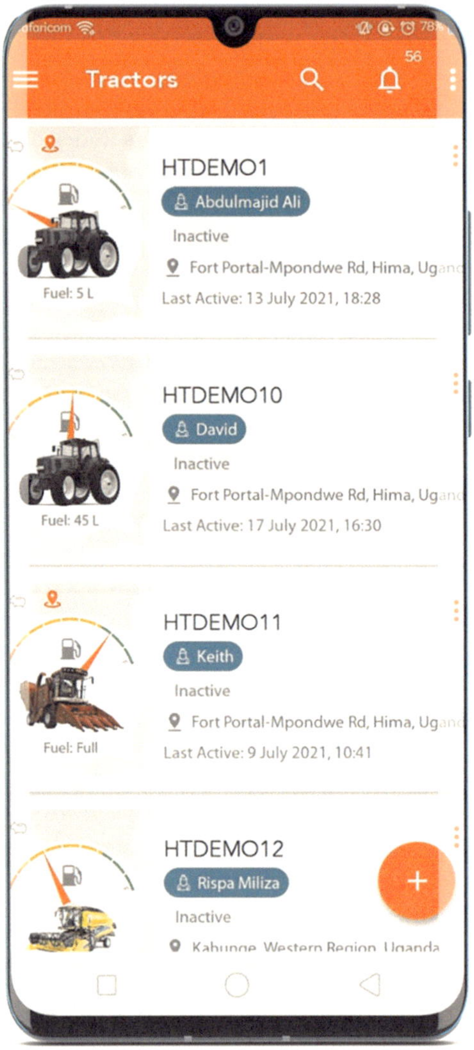

Figure 6.
The hello tractor application accessible to farmers via a smart phone.

This approach allows service providers to match seasonal demand for mechanization services and linked cash flows. Digital information and communication technology enable equipment owners to track the movement and use of their assets, expand their serviceable areas, and manage payment quickly and transparently. Reliable information and communication channels via smartphones allow clients equitable access to agricultural mechanization in ways that improve land productivity, reduce labor costs, and improve their incomes [40].

2.7 Pre-plant blended fertilizers and nitrogen topdressing

The right fertilizers must be applied at the right rate and at the right time, following best agronomic practices before smallholder maize producers across Africa can

Figure 7.
Dry rotary system used in small-scale fertilizer blending.

optimize grain yields. Shortages of soil nitrogen (N), phosphorus (P), and potassium (K) result in weak roots, stunted growth, greater vulnerability to pests and disease, reduced photosynthetic efficiency, fewer and smaller ears, and incomplete grain fill. Sub-Saharan Africa is facing food security challenges due, in part, to decades of soil fertility depletion. Applying mineral fertilizer, in conjunction with better management of organic resources and increasing Biological Nitrogen Fixation (BNF) can increase crop yields, replenish soil nutrients, increase soil organic carbon sequestration, and reduce N and C losses [41, 42]. Too few farmers in Sub-Saharan Africa use appropriate fertilizer formulations, dosages, and schedules, leading to lower yields, reduced profits, and nutrient-depleted soils [43].

Specialized blends of common fertilizers that contain N, P, K, and other nutrients such as sulfur, magnesium, and zinc are developed for basal application to maize crops. Applying blended fertilizers before planting can help to ensure a more balanced availability of nutrients for maize crops [44]. This is important because nitrogen fertilizer is one of the largest investments maize farmers make, and it can be lost due to drought or excessive rainfall. To overcome this inefficiency, it is widely recommended that nitrogen fertilizer be applied in two or more split applications throughout the growth cycle. This practice ensures that crops have a continuous supply of nitrogen, which can help mitigate financial risks to farmers and improve yields [45].

Many agro-dealers and manufacturers offer specially designed pre-plant fertilizer blends for maize. These formulations are adjusted to local growing conditions and soils and promote early crop development, stress resilience, and grain production by effectively delivering nutrients throughout the growing season. Top dressing N fertilizer later in the season better matches soil availability to the demand pattern of maize crops (**Figure 7**). The optimum time for top-dressing N fertilizer is when maize crops have eight to ten fully developed leaves. In this way, African farmers can obtain higher maize yields with lower rates of nutrient inputs when using blended fertilizers at planting instead of single fertilizers and splitting their nitrogen applications instead of a one-time input.

2.8 Maize-legume rotation and intercropping

Growing maize and grain legumes together as intercrops or in rotation offers many advantages compared to growing maize continuously as a monocrop [46]. Legumes increase nitrogen (N) in soils through biological nitrogen fixation and subsequent

Figure 8.
An innovative, staggered maize-soybean intercrop.

mineralization and can be used to offset the N requirements of the maize crop. Rotation and intercropping legumes with maize (**Figure 8**) improve the efficiency of land, nutrient, and water use due to synergistic effects between the crops [47]. Mixing maize and legumes also reduces the infestation of weeds, pests, and diseases in farmers' fields. Intercropping is crucial for small and marginal farmers in numerous countries as it diversifies and mitigates risks, improves the efficiency of land utilization, enhances soil fertility, and boosts economic returns, particularly in unpredictable weather conditions [48].

Large numbers of farmers in major maize production areas across Sub-Saharan Africa practice maize and legume rotation and intercropping, substantially increasing maize and legume yields and total harvests from a given land area. Growing a high-energy crop such as maize with high-protein legume results in improved diets among small-scale farmers and mitigates the risk of a hunger season when one of the two crops may fail because of drought or pest attacks. Biological nitrogen fixation in the root nodules of legumes benefits the productivity of maize crops rotated in the same field because part of the assimilated nitrogen is transferred between the crops through the decomposition of legume residues [49]. Mineral fertilizer application in mixed cropping systems is used very efficiently since either of the crops can benefit from residual nutrients that might have otherwise been lost due to the different root depths and distribution of maize and legumes [50]. Maize and legume intercropping are beneficial by reducing weed infestation, soil erosion, and run-off. This plant arrangement increases crop coverage and protection throughout the growing season.

Legumes can offer other advantages to maize crops. For example, soybeans and cowpeas can help to reduce parasitic striga weed infestations. This is because these legumes induce the germination of Striga seeds, but the weed does not infect them. Taller-statured maize benefits the legumes by better-regulating soil temperature soil through shading. However, understory legumes compete with maize for light, water, and nutrients. Intercropping can be a good way to increase maize yields and generate larger returns to labor. However, some challenges are associated with intercropping, such as careful crop selection and spacing. Additionally, some field operations, such as mechanization and chemical weeding, can be more difficult with intercropping systems [51].

2.9 Pre-emergent herbicides for weed management

Weeds can compromise maize croplands by competing for limited soil water and nutrients. Uncontrolled weeds can reduce yields and limit returns on agro-input investments. Controlling weeds in maize is critical, particularly during its early establishment and vegetative growth phases that extend to 10 weeks or so after planting. Without effective weed control, maize yields can be reduced by up to 50% on average [52], and losses can reach 80% if no measures are taken. In Africa, most smallholder farmers weed their maize crops by hand, a labor-intensive practice that must be repeated 2 or 3 times to be effective. This is because shallow hoeing can agitate the soil and promote the germination of weed seeds. Pre-emergent herbicides can help to reduce labor requirements by eliminating the need for hand weeding and help to improve soil quality by reducing the need for tillage, which can damage soil structure.

Pre-emergence herbicides prevent weeds from developing and allow fields to remain weed-free through the critical stages of crop establishment (**Figure 9**). This is important because it prevents weeds from competing with the maize crop for water, nutrients, and sunlight, which can help to reduce crop losses. This effect continues until the maize canopy shades the ground and weeds become suppressed [53]. This class of herbicides is applied shortly before or when planting maize and after the soil has been tilled. This technology prevents weed seedlings from establishing but requires that the proper chemicals are affordable, and that application equipment and safety gear are available. Some weeds emerge in most maize fields after crop establishment during the latter vegetative stage. These late-season weeds are effectively controlled by spraying recommended post-emergence herbicides to keep the fields

Figure 9.
Weedy (left) and weed-free maize understory.

clean until harvest, further enhancing maize productivity. Maize is very sensitive to competition from weeds between the emergence to the unfurling of six leaves. During this time, maize's fibrous root system is under development, and its shoots may become outcompeted by faster-growing plants. Maize gains the upper hand against weeds with pre-emergent herbicides. These herbicides remove the competition for light, nutrients, and moisture during maize's vulnerable initial growth phase. This, in turn, speeds up the growth of both roots and shoots.

2.10 New and emerging pest control practices

Fall Armyworm (FAW) invaded Africa in 2016 and continues to damage maize and many other crops (**Figure 10**). FAW is an aggressively damaging invader that afflicts the entire continent and affects numerous African crops [54]. Approximately US $13 billion worth of crops are at risk throughout Sub-Saharan Africa, threatening the livelihoods of many millions of smallholders [55]. FAW are the caterpillars of the invasive species *Spodoptera frugiperda*, and this destructive insect continues to spread across Sub-Saharan Africa [56]. Infestations of farmlands by the pest are caused by eggs deposited in soil and on the plants coming from adult moths that can fly and cover large distances. FAW larvae inflict extensive damage to maize crops at all life cycle stages by eating the whorl (apex), leaves, and ears, resulting in 50% yield loss or complete crop failure. Effective chemical control agents for FAW are known, but the pest has nonetheless spread across the continent and is threatening millions of farmers in major production zones.

A range of insecticide products are marketed on the continent by agro-input suppliers that kill larvae of FAW inside the soil and on the plant [57]. Coating maize seeds with insecticides protects the young maize plant from pest attack by enhancing seed survival, germination rates, and initial growth stages after planting [58]. Using insecticide as a seed treatment offers several advantages compared to foliar applications. The approach makes it possible to apply smaller amounts of the control agent and is positioned into the soil where eggs of FAW are deposited and hatched. FORTENZA® Duo seed coating technology from Syngenta has been demonstrated to be a powerful control agent for FAW and has been used to treat more than 3000 tons of maize seed in Zambia. Coating maize seeds is simple: mixing insecticide with a binding agent like gum Arabic, vaporizing it over the material, and letting it mix and

Figure 10.
Severe damage to maize inflicted by the fall armyworm that has recently invaded Africa.

dry in a rotary blending system. After treatment, the seed retains protective properties and provides a sufficient defense to the young seedling against FAW and other pests below and above the soil surface. Insecticides recommended for use as foliar spray later in the growing season are Ampligo® (chlorantraniliprole + lamba cyhalothrin), DenimFit® (emamectin benzoate+lufenuron), or Neconeem® (neem). It is vital to detect FAW infestations early so that control measures can be implemented before the pest causes too much damage.

2.11 Aflatoxin management

Common species of the soil-dwelling fungus *Aspergillus flavus* infest farmers' crops and foods, producing a highly toxic, cancer-causing poison called "aflatoxin" [59]. Widespread and severe contamination of several key staple crops, animal feeds, and processed foods occurs across Africa as a combined result of conducive weather conditions, extremely potent fungal strains, and substandard post-harvest handling and storage practices. In Africa, aflatoxin occurs not only in maize (**Figure 11**) and groundnut, where it poses a serious public health challenge, but also in cassava, sorghum, rice, and cashews, among others. When contaminated food is consumed by humans or livestock, aflatoxin accumulates inside the body and causes major damage to internal organs and blood. This toxin causes liver cancer, weakens people against other diseases, and stunts growth of children. Animals such as cows, pigs, and chickens are also affected by this toxin, and their milk, meat, and eggs become contaminated and unsafe for consumption. The aflatoxin pandemic in Africa has massive economic impacts by making food unfit to eat or trade, robbing humans of their health, and stunting and killing farm animals.

Biocontrol technologies for aflatoxin exist that rely upon natural competitors rather than industrial chemicals. These agents were safely and effectively adopted on increasingly large farmland areas over the past decade [60]. Aflasafe® is a product made in Africa that substantially reduces aflatoxin levels in food and is inexpensive and cost-effective to purchase and apply (**Figure 12**). The active ingredients of Aflasafe® are atoxic strains of *A. flavus* that do not produce the toxin. Combinations of four different strains are combined for each country by screening thousands of candidate strains recovered from local environments. Aflasafe® products are broadcast across crops 2 to 3 weeks before the onset of flowering. Alternatively, the product

Figure 11.
Infestation by A. flavus *causing maize to be unfit for consumption.*

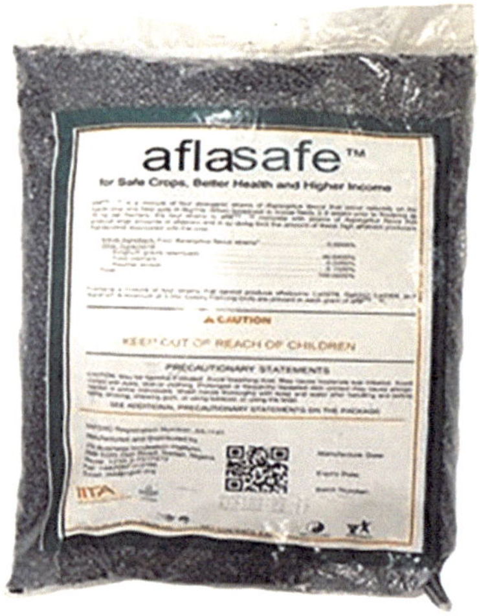

Figure 12.
Packaged Aflasafe, a product able to reduce the threat of mycotoxins.

may be applied onto the soil using a tractor-mounted spinner [61]. Different Aflasafe® products are produced and marketed in Burkina Faso, Ghana, Kenya, Malawi, Mozambique, Nigeria, Senegal, Tanzania, The Gambia, Uganda, and Zambia. Additional countries are in the process of identifying and registering biocontrol agents and constructing production facilities. Manufacturers of biological control technologies for aflatoxin must gain approval to use certified strains of atoxic fungi and comply with national regulations concerning the production, distribution, and release of microbial agents. Farmers do not require permits to apply Aflasafe® to their fields. The atoxic strains of *A. flavus* used in biocontrol are never copyrighted. However, they remain the genetic resources and property of the countries where they are developed for use as a public good. The IITA Business Incubation Platform is responsible for further developing and extending Aflasafe® across Sub-Saharan Africa.

2.12 Post-harvest management technologies

In Africa, post-harvest management technologies for maize focus on reducing losses and maintaining grain quality. These technologies include drying methods such as solar and mechanical dryers and improved traditional drying techniques. Storage precautions such as hermetic bags, metal silos, and plastic drums are utilized to protect maize from pests and moisture. Grain cleaning using mechanical grain cleaners or sieves helps remove impurities. Maize processing technologies, such as milling machines and dehullers, are employed to transform maize into different products. Integrated pest management techniques and quality testing tools ensure pest control and quality assurance. These technologies minimize post-harvest losses and improve the value of maize crops.

Hermetic bags are a type of storage technology with a three-layered design. The outer layer is made of woven polypropylene and provides the necessary strength to support the weight of the stored grain (**Figure 13**). Inside, there are two inner bags made of high-density polythene. These inner bags are specifically designed to have extremely low gas permeability and are water-resistant. The production process of

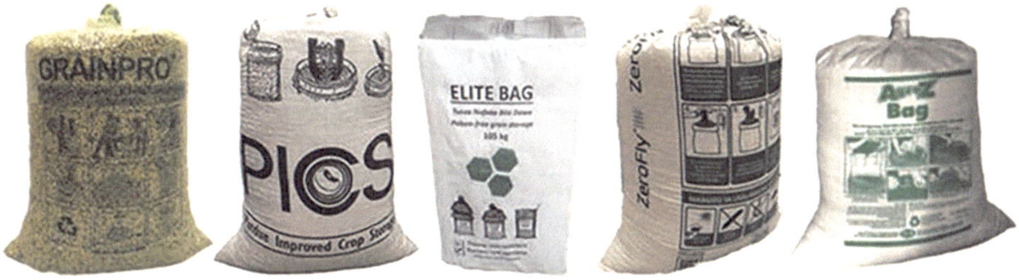

Figure 13.
Many different brands of hermetic grain storage bags are now available through agrodealers.

hermetic bags involves converting melted polypropylene into string form, which is then wound to create the polypropylene woven outer bag. A knitting machine weaves the string into the desired bag shape. For the polyethylene inner liners, recycled or raw plastic is melted and shaped into a thin layer, which is then cooled. The plastic is then combined into rolls and cut into the appropriate sizes for the inner bags. The purpose of hermetic bags is to create a barrier that prevents air and moisture from entering the stored grain. By cutting off the supply of oxygen, these bags effectively eliminate insects and microbial organisms, thus preserving the quality of the grain and reducing stored grain losses. They can provide storage for up to 2 years. Additionally, hermetic bags have gained cultural acceptance and are widely adopted by African farmers [62, 63].

Promoting hermetic bags prevents food loss and offers economic benefits to farmers and improved health outcomes due to reduced pesticide use and potential aflatoxin intake reduction. A study by Ndengwa et al. [64] found that compared to conventional methods, hermetic bags significantly curbed insect-related damage and weight loss with only 4% grain damage and 0.4% weight loss compared to the traditional practices group's 14% damage and 1.7% loss, over a crucial four-month storage period. The study also highlighted the potential profitability of hermetic bags with at least 4 months' seasonal usage across four seasons. Similarly, when produce quality is less crucial for a farmer's consumption, Dijkink et al. [65] reported that utilizing hermetic bags becomes economically advantageous compared to alternative storage methods for produce stored for more than 100 days.

3. Delivery of modernizing technologies

Developmental importance is attached to how proven, accompanying maize technologies are packaged for deployment and then managed as transferable assets within large programs and institutions [12, 66]. These technologies exist as production inputs, crop and land management options, and opportunities for contracted services. Combining these technologies into packages that result in improved yields offering reliable, profitable returns, and then scaling these packages to increasingly larger adopters may be viewed as central to agricultural transformation strategies, and major programs and institutional innovations are forming around this goal [11, 66–68]. In some cases, farmers are committed to older and traditional varieties for reasons other than their productive capacity or marketability, and efforts may be directed to convince them of a need for change [69].

3.1 Follow up to the Dakar 2 summit

The agricultural development community must mobilize and sustain country and development partners' commitment to agricultural transformation. To do so, Regional Member Countries of the African Development Bank first presented individual country food and Agriculture Delivery Compacts at the Dakar 2 Summit [70]. These planning documents are being formalized into standardized Agricultural Transformation Agendas through assistance from international development partners. Presidential Advisory Councils supervise each of the Country Compacts (see **Box 1**) led by the Head of State or their directly appointed representative and then report to the AfDB President through a Special Envoy. This mechanism is intended to provide high-level policy guidance toward the Feed Africa priorities. Several policies are associated with successful efforts toward agricultural transformation, including progressive regulation of seed systems, duty-free entry of agricultural inputs and equipment, ready movement of production inputs across borders, special incentives and provisions for agricultural loans, and others. Tracking the establishment and operations of the Country Compacts ensures that the necessary ingredients and actors needed for agricultural transformation are in place. The Dakar 2 process also involves working with key funding partners and the private sector to mobilize additional resources. The first challenge is to ensure that funds pledged for agricultural transformation materialize, and this is best accomplished by building confidence among different potential contributors that timely and significant progress is being made. In some cases, the Country Compacts represent a means to consolidate and more efficiently organize various,

> ACKNOWLEDGE that the Country Food and Agriculture Delivery Compacts developed at this Summit were prepared and are owned by African countries, which convey the vision, challenges, and opportunities in agricultural productivity, infrastructure, processing and value addition, markets, and financing that will accelerate the implementation of the African Union's Comprehensive Africa Agriculture Development Program (CAADP);
>
> AGREE that it is time for Africa to feed itself and fully unlock its agriculture potential to help feed the world;
>
> HEREBY RESOLVE to undertake the following:
>
> Finalize the development of the Country Food and Agriculture Delivery Compact endorsed at the Dakar 2 Summit in collaboration with country stakeholders, development partners, and the private sector to achieve food security and self-sufficiency;
>
> Establish Presidential Delivery Councils to oversee the implementation of the Country Food and Agriculture Delivery Compacts;
>
> Support the implementation of the Country Food and Agriculture Delivery Compacts with time-bound and clearly measurable indicators for success, including concrete national policies, incentives, and regulations to establish an enabling environment for wider and accelerated investments across the agriculture sector;
>
> Mobilize internal and external financing for the Country Food and Agriculture Delivery Compacts from a broad range of bilateral and multilateral partners and the private sector;
>
> Increase financing from national budgets to support the Country Food and Agriculture Delivery Compacts in line with the Malabo Declaration on Accelerated Agricultural Growth and Transformation for Shared Prosperity and Improved Livelihoods by allocating at least 10% of public expenditure to agriculture; and.
>
> Request that the African Union Commission and the African Development Bank follow up with various development partners to finalize their planned financial support to complement the $30 billion of financing announced at this Summit (now $70 billion) and to report on the overall investment of development partners; and ensure that the Dakar 2 Summit's Declaration is submitted to the February 2023 African Union Summit for consideration.

Box 1.
Declaration summary extracted from the Dakar 2 feed Africa summit.

and sometimes underperforming, agricultural development projects. Notably, underspending of past loans and grants because of disruption by the COVID-19 pandemic still occurs, and it is important to see these projects incorporated into and revitalized by the Country Compacts.

3.2 Emergence of the African Agricultural Leadership Institute (AALI)

AALI was formed shortly after the Dakar 2 event, led by the departure of the IITA Director General after 11 years of service. AALI's Strategy is embedded in a vision of establishing a new paradigm in the leadership of African agricultural development, resulting in accelerated agricultural sector modernization. AALI's agenda consists of three primary objectives (1) Provides advisory services to African governments seeking to modernize their agriculture and better implement their rural development agendas; (2) Empower youth as agricultural producers, service providers, and processors and restore agriculture as an attractive career path; and (3) Transforms agriculture through private sector growth resulting in the introduction of new technologies, needed production inputs and a next generation of service providers and agro-processors [71]. Achieving this agenda requires an enabling environment that helps countries expand agricultural growth through higher productivity on existing farmland, encourages strategic alliances within the continent, and revives the capacity for agricultural research and development through innovative problem-solving. Success requires that AALI operate an efficient internal organizational structure that guides the emergent Country Food and Agriculture Delivery Compacts emerging from the Dakar 2 Feed Africa Summit to establish precedents that guide agricultural transformation. Two of its foremost Objectives related to propelling the Dakar 2 Summit process forward relate to supporting African governments to develop innovative delivery mechanisms that translate vision and intent into concrete actions and benefits and guide current and future African political leaders and civil servants to acquire the leadership skills required to mobilize rural communities and to achieve pressing rural development agendas more successfully.

3.3 IITA's Partnerships for Delivery Directorate

The Partnerships for Delivery (P4D) Directorate aims to establish sustainable impact at scale and continues to expand rapidly in size and complexity. The Directorate operates under the authority of the IITA Board of Trustees and the supervision of the IITA Director General under the leadership of its Deputy Director General. It includes project and administrative support mechanisms provided to six Delivery Units: Development and Delivery, Youth in agribusiness, business incubation platform, mechanization, capacity development and communications. Each of these Units supports customized programs, projects, activities, and networks. P4D responded to two major opportunities: the unfolding of the One CGIAR agenda and the Dakar 2 Feed Africa Summit. The design of P4D is proving itself very strategic. Its Development and Delivery Unit is no longer a catchall for miscellaneous projects but has become a leader in Agricultural Transformation through its linkages to sovereign country loans and significant rural development efforts. The Youth in Agribusiness Unit [72] is no longer an exploratory curiosity but rather a platform for investment in the critical empowerment of young women and men through various approaches attractive to donors and national systems. With private partners, the P4D has proven to be a driving vehicle for increasing agricultural productivity by scaling technologies, promoting value chain development, and building economically sustainable seed systems. The Business Incubation Platform [73] has become the conveyor of proven technologies to the private sector while at the same time pivoting its orientation toward social enterprise in keeping with IITA's humanitarian principles. Maize is one of the focus commodities across all these efforts.

3.4 Technologies for African Agricultural Transformation (TAAT)

TAAT was launched in 2018 and renewed in 2022 through awards from the African Development Bank and the Bill and Melinda Gates Foundation. IITA is the executing agency of TAAT [12]. It is an integral component of AfDB's Feed Africa Strategy [11] and was well represented at the Dakar 2 Summit. TAAT ensures agricultural sector growth, improving food security and encouraging inclusive growth by involving more women and youth in profitable agricultural production and processing. Its larger goal is to improve agriculture as a business across Africa by deploying productivity-increasing agricultural technologies within nine priority food commodities: maize, cassava, wheat, rice, sorghum and millet, orange-fleshed sweet potato, high iron beans, aquaculture, and small livestock [12, 74]. By focusing efforts on these value chains, TAAT impacts agricultural productivity and diversification, leading to improved food and nutrition security, job creation, and agro-industrialization. Other benefits are reduced vulnerabilities to market price fluctuations due to more reliable supplies leading to better organized and accessible markets, improved soil, land and water management practices, and increased resilience to climate variability and stress. TAAT's technologies are described through a series of Technology Toolkit Catalogs, including one devoted to modernized maize production [51].

TAAT's Maize Technology Delivery Compact is mainly active in the savanna agro-ecosystems of East and Southern Africa. TAAT delivered water-efficient maize to 5.6 million households in Eastern Africa, an area hit by severe droughts. In Zambia, Zimbabwe, and Malawi, a TAAT-led collaboration with 15 private-sector seed companies reached 600,000 farmers with 6000 MT of drought-tolerant maize varieties (see Section 2.1) treated with specialized dual-purpose pesticides with demonstrated capabilities to control Fall Armyworm (see Section 2.10). TAAT promotes seed treatment with Fortenza Duo (FD) to combat invasive Fall Armyworm (FAW). In Zambia and Zimbabwe, TAAT deployed 6598 metric tons of certified maize seed treated with Fortenza Duo through government programs and reached 660,000 beneficiaries. An internal report of an impact study commissioned by TAAT found a 1.5 MT/ha yield improvement among farmers who used the FD-treated seeds compared to those who did not.

3.5 DR Congo Agricultural Transformation Agenda

The Agenda for the Transformation of Agriculture in the Democratic Republic of Congo (ATA-DRC) is fulfilling a Presidential promise to modernize agriculture [68]. The Government appointed IITA to lead this Agenda in early 2022. While it has a nationwide mandate, its first phase commenced in five carefully selected locations, focusing on maize, beans, soybeans, cassava, rice, banana, and aquaculture, the first three of which are particularly important within maize-based systems. It increases agricultural production by using improved crop varieties and building a solid seed system in close collaboration with the national agriculture research system and regulatory bodies. In addition, ATA-DRC provides other production inputs and good agricultural practices and adds value at the community level by engaging private sector operators in ways that build agro-industrial capacity and reduce food imports. Its goal is to create wealth and jobs through modernized agriculture by consolidating and building upon IITA expertise and several ongoing and planned future development projects. IITA works closely with Bio Agronomic Business (BAB), appointed as a national counterpart by the Ministry of Agriculture, with initial attention focused on realizing the potential of large state farms in different parts of the country.

In its short lifespan, ATA-DRC has produced some remarkable results. Starting with the 2022–2023 growing season, this program established 1518 ha into modernized crop production, including 547 ha of maize, 864 ha of cassava, and 81 ha of soybean. Most of this area is on previously underperforming state farms (83%) but with increasing attention on establishing vibrant out-grower networks. Seed production occurs on an additional 434 ha, including 188 ha of IITA's improved cassava varieties, soon destined to provide about 38 million cuttings. To date, 979 tons of improved maize, soybean, bean, and rice seeds have been produced for distribution to national partners. IITA's Semi Autotrophic Hydroponics (SAH) Technology is producing improved, disease-free cassava plantlets in two locations and is drawing investors to expand the SAH technology to other sites. IITA expertise is applied to existing cassava processing facilities, increasing production of High-Quality Cassava Flour from negligible to 4 tons per hour. This engineering expertise is also used in the milling of grains and will be applied to the production of animal feeds and biogas. Organizing the "Brigade du Pain" allows cassava flour to substitute for imported wheat flour across hundreds of bakeries partially. Over 100,000 fish fingerlings were produced in Kinshasa to promote aquaculture, and 10 tons of Aflasafe were made at the IITA Kalambo factory to spearhead food safety (see Section 2.11). DRC-ATA has put in place the essential building blocks to create impact at scale in the short run, including improvements in the maize value chain.

The agenda is charting a proven pathway to modernized agriculture across DRC in close collaboration with its national counterparts and private sector operators. It works with a Special Advisor to the President and even consults directly with H.E. Felix Antoine Tshisekedi. Seed systems gains are moving toward private and community-based seed producers. The production and processing facilities at the state farms are unlocking the great potential to serve as the models forpublic–private partnerships, demonstrating the profitability of agro-processing to lure further private-sector investment. Out-grower networks are forming around these facilities to ensure an adequate and reliable supply of raw materials and access to steady markets. IITA also partners with the African Agricultural Leadership Institute at the national level to engage in promoting a conducive policy environment and, at the field level, has been instrumental in establishing a nationwide "Brigade des Jeunes" (Youth Brigade) based in part upon many of the approaches of the IITA Youth Agripreneurs [72]. Farm mechanization is essential for scaling up operations, and the Brigadiers have introduced small-scale fields and processing equipment to help achieve this. Most importantly, DRC-ATA serves as an example for scaling operations to be replicated by the Dakar 2 process and its Country Compacts, starting with efforts in DR Congo.

4. Conclusions

This chapter provides a short history of maize in Africa, including its importance as a staple food, and identifies various accompanying technologies for modernizing maize production. It then describes some unfolding mechanisms to deploy these technologies within larger development thrusts. The chapter features high-yielding varieties that resist drought and pests and those that improve their nutritional value. It provides opportunities for supplying improved maize seed through recent mechanisms for commercial licensing and access to mechanized agricultural equipment and contracted services through digital agriculture platforms. It highlights fertilizer and soil nutrient management advances in maize-based systems, including pre-plant and top-dress fertilizers and legumes, to increase soil nitrogen stocks. Advances in weed management include the use of specialty and pre-emergent herbicides. It also provides insights into the control of invasive Fall Armyworms. It further features biotechnology that prevents aflatoxin contaminants from entering food systems. Maize grain is an important human food, but it can also be processed into high-quality flour and starches from which various products are manufactured. In addition, maize stover is widely used as fodder for livestock and important for practices like mulching and the maintenance of soil organic matter. Technologies featured in this Chapter offer the means for farming communities in Africa to access the high-end of the maize value chain and its global marketplace, improving returns to both small-scale farmers and commercial agribusinesses. The Feed Africa Strategy of the African Development Bank, the partnership galvanized around that Strategy, and the momentum achieved through the recent Dakar 2 Summit are viewed as promising means to deploy these technologies.

The authors note with concern that The Democratic Republic of Congo (DRC) is currently experiencing a maize crisis because national demand now far exceeds domestic supplies. Its government seeks a combined federal, international, and private sector response. Maize production for the DRC has grown from 306,000 MT in 1971 to 2 million MT in 2020, increasing at a rate of 4.5% per year. This growth was caused more by expanding land under cultivation rather than improving maize productivity. Land area under maize cultivation increased from 1.5 million ha in 2001 to 2.9 million ha in 2021, but maize yields remain low, averaging only 0.8 MT per ha. As a result, maize deficits of about 2.8 million MT developed a shortage that was largely addressed through importation from Zambia. But Zambia recently halted maize exports to cope with its own domestic shortages. As a result, the cost of maize flour on RDC has skyrocketed, increasing in some parts of the country from US $0.45 per kg a few months previously to $1.61 per kg in May 2023. A recent communication from the Deputy Prime Minister in charge of the economy stated, "The causes of this situation include the shortfall in local production in line with demand, restrictions on Zambian exports and high import costs, as well as the deterioration of climatic conditions, which affects agricultural production in the sub-region". Recent outreach efforts by ATA-DRC providing farming communities in Kasai and Lualaba with improved maize management practices resulted in yields of 1.7 MT per ha, a readily achieved increase of 112%. More concentrated efforts relying upon improved varieties, judiciously applied pre-plant and top-dressed fertilizer, better weed control, and other technologies described in this Chapter readily achieve 3 MT per ha yields. In this way, maize production in DRC may be improved by 2.6 to 6.4 million tons per year provided technologies described in this Chapter are scaled through increasingly available agricultural transformation processes.

Acknowledgements

The authors appreciate the valuable comments from Issac Balume and Jan A. Helsen of IITA's DRC Agricultural Transformation Agenda, and Nteranya Sanginga of the African Agricultural Leadership Institute. Parts of this Chapter rely upon data downloaded from the Product Platform for Agricultural Solutions (ProPAS) internet portal (see https://propas.iita.org). Photographs supporting technology descriptions are provided by the TAAT Clearinghouse and other projects led by IITA. Preparation of this Chapter was supported through the IITA Partnerships for Delivery Directorate.

Author details

Paul L. Woomer*, Dries Roobroeck and Welissa Mulei
International Institute of Tropical Agriculture, Nairobi, Kenya

*Address all correspondence to: w.mulei@cgiar.org

References

[1] Cherniwchan J, Moreno-Cruz J. Maize and precolonial Africa. Journal of Development Economics. 2019;**136**: 137-150

[2] Miracle MP. The introduction and spread of maize in Africa. The Journal of African History. 1965;**6**(1):39-55

[3] Purseglove JW. Tropical Crops: Monocotyledons. London: Longman Group Limited; 1972. p. 607

[4] McCann JC. Maize and Grace: Africa's Encounter with a New World Crop. Cambridge: Harvard University Press; 2007. p. 289

[5] Erenstein O, Jaleta M, Sonder K, Mottaleb K, Prasanna BM. Global maize production, consumption and trade: Trends and R&D implications. Food Security. 2022;**14**:1295-1319. DOI: 10.1007/s12571-022-01288-7

[6] Van Dijk M, Morley T, Jongeneel R, van Ittersum M, Reidsma P, Ruben R. Disentangling agronomic and economic yield gaps: An integrated framework and application. Agricultural Systems. 2017; **154**:90-99

[7] Cairns J et al. Challenges for sustainable maize production of smallholder farmers in sub-Saharan Africa. Journal of Cereal Science. 2021;**101**:103274. DOI: 10.1016/J. JCS.2021.103274

[8] Shoko RR et al. Estimating the supply response of maize in South Africa: A Nerlovian partial adjustment model approach. Agrekon. 2016;**55**:237-253. DOI: 10.1080/03031853.2016.1203802

[9] Shikur ZH et al. Agricultural policies, agricultural production and rural households' welfare in Ethiopia. Journal of Economic Structures. 2020;**9**:1-21. DOI: 10.1186/s40008-020-00228-y

[10] Byerlee D, Heisey P. Past and potential impacts of maize research in sub-Saharan Africa: A critical assessment. Food Policy. 1996;**21**:255-277. DOI: 10.1016/ 0306-9192(95)00076-3

[11] African Development Bank (AfDB). Feed Africa: Strategy for Agricultural Transformation in Africa 2016–2025. Abidjan, Cote d'Ivoire: AfDB; 2016. p. 79

[12] Woomer PL, Mulei WM, Zozo R. A new paradigm in the delivery of modernizing agricultural technologies across Africa. In: Technology in Agriculture. London: Intech Open; 2021. p. 23. DOI: 10.5772/intechopen.98940

[13] Lunduka RW, Mateva KI, Magorokosho C, Manjeru P. Impact of adoption of drought-tolerant maize varieties on total maize production in southeastern Zimbabwe. Climate and Development. 2019;**11**:35-46. DOI: 10.1080/17565529.2017.1372269

[14] Simtowe F, Amondo E, Marenya P, Rahut D, Sonder K, Erenstein O. Impacts of drought-tolerant maize varieties on productivity, risk, and resource use: Evidence from Uganda. Land Use Policy. 2019;**88**:104091. DOI: 10.1016/j. landusepol.2019.104091

[15] Chemura A, Nangombe SS, Gleixner S, Chinyoka S, Gornott C. Changes in climate extremes and their effect on maize (Zea mays L.) suitability over southern Africa. Frontiers in Climate. 2022;**4**:890210. DOI: 10.3389/ fclim.2022.890210

[16] MacRobert J, Setimela P, Gethi J, Worku RM. Maize Hybrid Seed

Production Manual. Mexico City, Mexico: CIMMYT. 2014. p. 26

[17] Obunyali C, Karanja J, Oikeh S, Omanya G, Mugo S, Beyene Y, et al. On-farm performance and farmers' perceptions of drought TEGO -climate-smart maize hybrids in Kenya. Agronomy Journal. 2019;**111**(6):2754-2768. DOI: 10.2134/agronj2019.08.0600

[18] Oniang'o RK, Obunyali CO, Oikeh SO. Adoption of climate-smart DroughtTEGO® varieties in Kenya. African Journal of Food, Agriculture, Nutrition & Development. 2019;**19**(4): 15090-15108

[19] Atera EA, Ishii T, Onyango JC, Itoh K, Azuma T. Striga infestation in Kenya: Status, distribution and management options. Sustainable Agriculture Research. 2013;**2019**(2): 99-108. DOI: 10.5539/sar.v2n2p99

[20] Badu-Apraku B, Fakorede MA, Akinwale RO, Adewale SA, Akaogu IC. Developing high-yielding Striga-resistant maize in sub-Saharan Africa. CABI Reviews. 2021;**16**(030):1-12. DOI: 10.1079/PAVSNNR202116030

[21] Woomer PL, Bokanga M, Odhiambo GD. Striga management and the African farmer. Outlook on Agriculture. 2008;**37**:277-282

[22] De Groote H, Wangare L, Kanampiu F. Evaluating the use of herbicide-coated imidazolinone-resistant (IR) maize seeds to control Striga in farmers' fields in Kenya. Crop Protection. 2007;**26**:1496-1506. DOI: 10.1016/j.cropro.2006.12.013

[23] Sibuti MS, Paul K, Joshua O. Effectiveness of Imazapyr coated hybrids and selected Striga-tolerant varieties on S. hermonthica management and maize yield performance in Western part of Kenya. Advances in Applied Physiology. 2021;**6**(1):1-8. DOI: 10.11648/J.AAP.20210601.11

[24] Kamara A, Ellis-Jones J, Amaza P, Omoigui L, Helsen J, Dugje I, et al. A participatory approach to increasing productivity of maize through Striga Hermonthica control in Northeast Nigeria. Experimental Agriculture. 2008; **44**:349-364. DOI: 10.1017/S0014479708006418

[25] Kanampiu F, Makumbi D, Mageto E, Omanya G, Waruingi S, Musyoka P, et al. Assessment of management options on Striga infestation and maize grain yield in Kenya. Weed Science. 2018;**66**: 516-524

[26] Mignouna DB, Manyong VM, Mutabazi KD, Senkondo EM. Determinants of Adopting Imazapyr-Resistant Maize for Striga Control in Western Kenya: A Double-Hurdle Approach. Tanzania: Sokoine University of Agriculture; 2011. Available from: https://www.suaire.sua.ac.tz/handle/123456789/4036

[27] CIMMYT. Biofortified maize and wheat can improve diets and health; new study shows. 2019. Available from: https://www.cimmyt.org/news/biofortified-maize-and-wheat-can-improve-diets-and-health-new-study-shows/

[28] Ewool MB, Akromah R, Acheampong PP. Performance of pro-vitamin a maize synthetics and hybrids selected for release in Ghana. International. Journal of Science and Technology. 2016;**5**:268-291

[29] TAAT Program. Technology Pitch: Golden Maize Varieties (Vitamin A Biofortified). 2023a. Available from: https://knowledgecenter.taat-africa.

org/wp-content/uploads/2023/02/Golden-maize-varieties_EN.pdf

[30] Nesamvuni A, Vorster H, Margetts B, Kruger A. Fortification of maize meal improved the nutritional status of 1–3-year-old African children. Public Health Nutrition. 2005;**8**:461-467. DOI: 10.1079/PHN2005782

[31] Raviya P, Savaliya V, Vaghasiya K, Gohil G. Knowledge level of farmers regarding information and communication technology services. International Journal of Current Microbiology and Applied Sciences. 2020. DOI: 10.20546/ijcmas.2020.908.110

[32] Ayim C, Kassahun A, Addison C, Tekinerdogan B. Adoption of ICT innovations in the agriculture sector in Africa: A review of the literature. Agriculture & Food Security. 2022;**11**:22. DOI: 10.1186/s40066-022-00364-7

[33] Maumbe B, Okello J. Uses of information and communication technology (ICT) in agriculture and rural development in sub-Saharan Africa: Experiences from South Africa and Kenya. International Journal of ICT Research and Development in Africa. 2010;**1**:1-22. DOI: 10.4018/jictrda.2010010101

[34] AATF. Championing agricultural technology transfer in Africa. Annual Report 2021. African Agricultural Technology Foundation (AATF), Nairobi, Kenya. 2022

[35] George M, Ibrahim M, Grace M, Stephen M, Ruth R, Ruth KO, et al. Impact of DroughtTEGO hybrid maize variety on agricultural productivity and poverty alleviation in Kenya. African Journal of Agricultural Research. 2019;**14**(34):1833-1844. DOI: 10.5897/AJAR2019.14237

[36] Zaidi PH, Vinayan MT, Nair SK, Kuchanur PH, Kumar R, Singh SB, et al. Heat-tolerant maize for rainfed hot, dry environments in the lowland tropics: From breeding to improved seed delivery. The Crop Journal. 2023;**11**(4):986-1000. DOI: 10.1016/j.cj.2023.06.008

[37] Anidi O. Use of Information and Communications Technology Tools for Tractor Hire Services in Africa. Rome, Italy: FAO; 2023 Available from: https://policycommons.net/artifacts/1421808/use-of-information-and-communications-technology-tools-for-tractor-hire-services-in-africa/2035859/

[38] Kirui O. The Agricultural Mechanization in Africa: Micro-Level Analysis of State Drivers and Effects. Econometric Modeling: Agriculture. University of Bonn - Center for Development Research (ZEF); 2019. DOI: 10.2139/ssrn.3368103

[39] Daum T, Adegbola PY, Adegbola C, Daudu C, Issa F, Kamau G, et al. Mechanization, digitalization, and rural youth-stakeholder perceptions on three mega-topics for agricultural transformation in four African countries. Global Food Security. 2022;**32**:100616 Available from: https://www.sciencedirect.com/science/article/pii/S2211912422000074

[40] The Borgen Project. Uber for Tractors: Transforming the Agricultural Sector in Africa. 2023. Available from: https://borgenproject.org/the-agricultural-sector-in-africa/

[41] Wei Z, Ying H, Guo X, Zhuang M, Cui Z, Zhang F. Substitution of mineral fertilizer with organic fertilizer in maize systems: A meta-analysis of reduced nitrogen and carbon emissions. Agronomy. 2020;**10**:14. DOI: 10.3390/agronomy10081149

[42] MacCarthy D, Vlek P, Bationo A, Tabo R, Fosu M. Modeling nutrient and water productivity of sorghum in smallholder farming systems in a semi-arid region of Ghana. Field Crops Research. 2010;**118**:251-258. DOI: 10.1016/J.FCR.2010.06.005

[43] Pasley HR, Camberato JJ, Cairns JE, et al. Nitrogen rate impacts on tropical maize nitrogen use efficiency and soil nitrogen depletion in eastern and southern Africa. Nutrient Cycling in Agroecosystems. 2020;**116**:397-408. DOI: 10.1007/s10705-020-10049-x

[44] Zingore S, Njoroge S, Chikowo R, Kihara J, Nziguheba G, Nyamangara J. 4R Plant Nutrient Management in African Agriculture: An Extension Handbook for Fertilizer Management in Smallholder Farming Systems. IFDC: Nairobi, Kenya; 2014. p. 90

[45] Mosisa W, Dechassa N, Kibret K, Zeleke H, Bekeko Z. Effects of timing and nitrogen fertilizer application rates on maize yield components and yield in eastern Ethiopia. Agrosystems, Geosciences and Environment. 2022;**5**: 20322. DOI: 10.1002/agg2.20322

[46] Baijukya F, Wairegi L, Giller K, Zingore S, Chikowo R, Mapfumo P. Maize-legume cropping guide. Nairobi, Kenya: Africa Soil Health Consortium; 2016. p. 88

[47] Bationo A, Chianu J, Giller K, Idupulapati R, Lesueur D, Merckx R, et al. Integrated soil fertility management in Africa: Principles, practices and developmental process. In: Sanginga N, Woomer PL, editors. Tropical Soil Biology and Fertility. Nairobi: Institute of the International Centre for Tropical Agriculture; 2009. p. 263

[48] Tripathi S, Venkatesh K, Meena R, Chander S, Singh G. Sustainable intensification of maize and wheat cropping system through pulse intercropping. Scientific Reports. 2021; **11**:10. DOI: 10.1038/s41598-021-98179-2

[49] Woomer PL. Biological Nitrogen Fixation and Grain Legume Enterprise: Guidelines for N2Africa Master Farmers. Nairobi: Tropical Soil Biology and Fertility Institute of the International Centre for Tropical Agriculture; 2010. p. 17

[50] Ngwira AR, Aune JB, Mkwinda S. On-farm evaluation of yield and economic benefit of short term maize legume intercropping systems under conservation agriculture in Malawi. Field Crops Research. 2012;**132**:149-157. DOI: 10.1016/j.fcr.2011.12.014

[51] Maitra S, Shankar T, Banerjee P. Potential and advantages of maize-legume intercropping system. In: Maize-Production and Use. London: IntechOpen; 2020. DOI: 10.5772/intechopen.91722

[52] Chikoye D, Lum AF, Abaidoo R, Menkir A, Kamara A, Ekeleme F, et al. Response of corn genotypes to weed interference and nitrogen in Nigeria. Weed Science. 2008;**56**:424-433. DOI: 10.1614/WS-07-055.1

[53] Imoloame EO. Evaluation of herbicide mixtures and manual weed control method in maize (Zea mays L.) production in the southern Guinea agro-ecology of Nigeria. Cogent Food & Agriculture. 2017;**3**:1375378. DOI: 10.1080/23311932.2017.1375378

[54] Kansiime MK, Rwomushana I, Mugambi I. Fall armyworm invasion in sub-Saharan Africa and impacts on community sustainability in the wake of coronavirus disease 2019: Reviewing the evidence. Current Opinion in

Environmental Sustainability. 2023;**62**: 101279. DOI: 10.1016/j.cosust.2023.101279

[55] Harrison RD, Thierfelder C, Baudron F, Chinwada P, Midega C, Schaffner U, et al. Agro-ecological options for fall armyworm (Spodoptera frugiperda JE smith) management: Providing low-cost, smallholder friendly solutions to an invasive pest. Journal of Environmental Management. 2019;**243**: 318-330

[56] FAO and CABI. Fall Armyworm Field Handbook: Identification and Management. First ed. FAO: Rome, Italy; 2019. p. 36

[57] TAAT Clearinghouse. Maize Technology Toolkit Catalogue, Clearinghouse Technical Report Series 008, Technologies for African Agricultural Transformation. IITA, Cotonou, Benin: Clearinghouse Office; 2021. p. 31

[58] Han H, Chen B, Xu H, Qin Y, Wang G, Lv Z, et al. Control of Spodoptera frugiperda on fresh corn via pesticide application before transplanting. Agriculture. 2023;**13**:342. DOI: 10.3390/agriculture13020342

[59] Atehnkeng J, Donner M, Ojiambo PS, Ikotun B, Augusto J, Cotty PJ, et al. Environmental distribution and genetic diversity of vegetative compatibility groups determine biocontrol strategies to mitigate aflatoxin contamination of maize by *Aspergillus flavus*. Microbial Biotechnology. 2016;**9**:75-88. DOI: 10.1111/1751-7915.12324

[60] Ola OT, Ogedengbe OO, Raji TM, Eze B, Chama M, Ilori ON, et al. Aflatoxin biocontrol effectiveness in the real world: Private sector-led efforts to manage aflatoxins in Nigeria through biocontrol-centered strategies. Frontiers in Microbiology. 2022;**13**:977789. DOI: 10.3389/fmicb.2022.977789

[61] Atehnkeng J, Mutegi C, Ortega-Beltran A, Augusto J, Akande A, Senghor L, et al. Management of aflatoxins in maize in Africa: Trade-off between reducing exposure and avoiding increased risk of contamination. World Mycotoxin Journal. 2018;**11**:173-190. DOI: 10.3920/WMJ2017.2317

[62] TAAT Program. Technology Pitch: Hermatic Bags for Safe Storage of Grains. 2023b. Available from: https://knowledgecenter.taat-africa.org/wp-content/uploads/2023/02/Hermetics-bags_resellers_EN.pdf

[63] Mlambo S, Mvumi B, Stathers T, Mubayiwa M, Nyabako T. Field efficacy of hermetic and other maize grain storage options under smallholder farmer management. Crop Protection. 2017;**98**:198-210. DOI: 10.1016/J.CROPRO.2017.04.001

[64] Ndegwa M, Groote H, Gitonga Z, Bruce A. Effectiveness and economics of hermetic bags for maize storage: Results of a randomized controlled trial in Kenya. Crop Protection. 2016;**90**:17-26. DOI: 10.1016/J.CROPRO.2016.08.007

[65] Dijkink B, Broeze J, Vollebregt M. Hermetic bags for the storage of maize: Perspectives on economics. Food Security and Greenhouse Gas Emissions in Different Sub-Saharan African Countries. 2022;**6**. Working Paper No. 291. DOI: 10.3389/fsufs.2022.767089

[66] Barrett CB, Benton TG, Fanzo J, Herrero M, Nelson RJ, Bageant E, et al. Socio-Technical Innovation Bundles for Agri-Food Systems Transformation, Report of the International Expert Panel on Innovations to Build Sustainable, Equitable, Inclusive Food Value Chains

[Internet]. Ithaca, NY, and London: Cornell Atkinson Center for Sustainability and Springer Nature; 2020. p. 172 Available from: https://hdl.handle.net/10568/110864

[67] Sartas M, Schut M, Proietti C, Thiele G, Leeuwis C. Scaling readiness: Science and practice of an approach to enhance impact of research for development. Agricultural Systems. 2020;**184**:12. DOI: 10.1016/j.agsy.2020.102874

[68] Woomer PL, Zozo RM, Lewis S, Roobroeck D. Technology promotion and scaling in support of commodity value chain development in Africa. In: Stanton J, editor. Agricultural Value Chains - some Selected Issues. London, UK: IntechOpen; 2023. pp. 1-29. DOI: 10.5772/intechopen.110397

[69] Abate T, Fisher M, Abdoulaye T, et al. Characteristics of maize cultivars in Africa: How modern are they and how many do smallholder farmers grow? Agriculture & Food Security. 2017;**6**:30. DOI: 10.1186/s40066-017-0108-6

[70] Akinbamijo Y. Africa Food Summit Dakar 2: Africa's Agrifood System and Triple Helix. Accra, Ghana: Forum for Agricultural Research in Africa; 2023. Available from: https://faraafrica.org/2023/02/09/africa-food-summit-dakar-2-africas-agrifood-system-and-triple-helix/

[71] Sanginga N. African Agricultural Leadership Institute: Strategy 2023–2033. Bukavu, République Démocratique du Congo: African Agricultural Leadership Institute (AALI); 2023. p. 27

[72] Sanginga N, Adenmosun A, Obaniyi J, Mulinganya N, Woomer P. The IITA agripreneur movement: A dynamic approach to youth empowerment across Africa. Journal of International Agricultural and Extension Education. 2023;**30**(2):82-115. DOI: 10.4148/2831-5960.1059

[73] IITA-BIP. IITA Business Incubation Platform. Ibadan, Nigeria: International Institute of Tropical Agriculture; 2023 Available from: https://iitabip.com/

[74] TAAT Clearinghouse. TAAT technology toolkits and their strategic deployment. Clearinghouse [technical report] series 001, Technologies for African Agricultural transformation, Clearinghouse Office, Cotonou, Benin. 2018. p. 18

Section 4

Maize Breeding and Genomics

Chapter 7

Enhancing Maize (*Zea mays* L.) Crop through Advanced Techniques: A Comprehensive Approach

Vikas Kumar Singh, Saksham Pundir, Deepti Chaturvedi, Arshdeep Kaur, Abhishek Pandey, Shreya Mandal, Rahul Kumar, Ramesh Kumar Singh, Harish Pal Bhati, Parmdeep Singh Dhanda, Alpa Yadav, Chittaranjan Kole and Prashant Kaushik

Abstract

Maize (*Zea mays* L.) is one of the most widely cultivated crops globally, making significant contributions to food, animal feed, and biofuel production. However, maize yield is greatly affected by various climate and soil factors, and it faces hindrances due to abiotic stresses, such as drought, salinity, extreme temperatures, and cold conditions. In confronting these hurdles, the field of crop breeding has transformed thanks to high-throughput sequencing technologies (HSTs). These advancements have streamlined the identification of beneficial quantitative trait loci (QTL), associations between markers and traits (MTAs), as well as genes and alleles that contribute to crop improvement. Presently, well-established omics techniques like genomics, transcriptomics, proteomics, and metabolomics are being integrated into maize breeding studies. These approaches have unveiled new biological markers can enhance maize's ability to withstand a range of challenges. In this chapter, we explore the current understanding of the morpho-physiological and molecular mechanisms underlying maize resistance and tolerance to biotic and abiotic stresses. We focus on the use of omics techniques to enhance maize's ability to withstand these challenges. Moreover, it emphasizes the significant potential of integrating multiple omics techniques to tackle the challenges presented by biotic and abiotic stress in maize productivity, contrasting with singular approaches.

Keywords: maize (*Zea mays* L.), omics, stresses, resistance, crop improvement

1. Introduction

Maize (*Zea mays* L.), commonly referred to as maize holds a prominent and cherished status as one of the world's most essential crops. Its importance extends beyond geographical borders, influencing cultures, economies, and dietary habits worldwide. Maize, serving as a staple food source for humans, providing nourishment, supporting livelihoods, a fermentation substrate, and a valuable commodity in numerous industrial applications [1], particularly in its dry grain form, occupies a critical role on the global scale. Wheat, maize, and rice are the primary staple cereals worldwide, each cultivated on approximately 200 million hectares. Corn, frequently referred to as maize, underwent domestication more than 9000 years ago within the southern Mexico/Mesoamerica region [2]. Together, these three primary global staple cereals, namely wheat, rice, and maize, make up a substantial portion of the human diet, accounting for approximately 40 percent of the world's calorie intake and 35% of protein consumption [3]. Maize fulfills a versatile and continually evolving function within the global agricultural and food systems, making substantial contributions to food and nutrition security [4, 5]. About 56% of its output is used as livestock feed, while one-fifth finds application in non-food sectors, and 13% is designated for human consumption. Notably, maize is distinguished by its high starch content, constituting roughly 65% of its composition [6]. Currently, maize has risen as a viable alternative to rice and wheat. Around 35% of its harvest is directed toward human consumption, while 25% serves as feed for poultry and cattle, and another 15% is used in food processing [1]. It has achieved the status of a significant global commodity, with 15% of the world's maize production currently being exported, marking an increase from the 11% reported in the previous decade [3]. It is on the verge of overtaking wheat as the most heavily traded cereal. Leading net-exporting nations such as the USA, Brazil, Argentina, Ukraine, and Romania are consistently shipping substantial quantities, ranging from 5 to 54 million metric tons annually [7]. Over the last century, maize yield has surged by a factor of eight, thanks to innovations in yield per plant and plant density optimization achieved through harnessing heterosis.

Throughout history, maize has been a quintessential subject in the realms of genetics, developmental biology, physiology, and, more recently, genomic research. The genetic investigation of *Zea mays* L. commenced with Edward East's pioneering research in 1908, which explored topics like inbreeding depression and hybrid vigor. A significant advancement in cytogenetics occurred in the 1940s when transposable elements (TEs) were discovered, as exemplified by Barbara McClintock's pioneering work [8]. In 2009, the first maize genome was made public [9]. Then, Jiao et al. [10] embarked on a re-sequencing project focused on the B73 maize variety. This effort revealed that a substantial 74% of its genome is comprised of long-terminal repeat retrotransposons (LTRs), which predominantly contribute to its enlarged size in comparison to other grass species. The main factor responsible for the maize genome's expansion relative to other grasses is the widespread increase in LTRs [10]. The cumulative cytogenetic, genetic, and genomic studies of maize have yielded rich insights into its genome's structure, function, and evolution. Resequencing wild relatives, traditional landraces, and improved maize lines, and aligning them with the reference genome, suggests that introgression from wild relatives contributes to post-domestication maize diversity. Through this method, genes with a wide range of biological functions that experienced selective pressure during the domestication process have been pinpointed [11].

The foundation of maize breeding relies on leveraging heterosis, which involves genome-wide allelic interactions, interactions among quantitative trait loci (QTLs), and inter-genomic interactions that occur when the two parental genomes combine in the F1 hybrid. The functional understanding of many maize genes, especially those linked to heterosis, is less advanced compared to other cereal crops, notably wheat and rice. Integrating whole-genome markers into genomic-based breeding represents a viable approach for improving maize breeding and holds significant promise. Genomic selection (GS) serves as a notable illustration of this genomic design breeding strategy as it does not require an extensive comprehension of gene functions or the precise assessment of each marker's efficacy [12]. In this contemporary genomics era, the integration of various strategies and methodologies promises to boost maize productivity. This includes the amalgamation of modern genomics, phenomics, gene editing, synthetic biology, and the utilization of AI technology. The integration of an extensive array of maize omics data, spanning genomics, phenomics, epigenomics, transcriptomics, proteomics, and metabolomics, will form a vital foundation for machine learning approaches to build network models illustrating the interactions among various genetic components. In the context of a maize breeding program, a key strategic approach for achieving the desired goals of increased production and enhanced quality traits revolves around reducing the frequency of hybridizations while maximizing the incorporation of superior alleles. Swift progress toward trait enhancement objectives can be attained by executing two or three iterations of small-scale population development, thereby making the most of the available genetic diversity. This chapter highlights the importance of incorporating comprehensive strategies to enhance maize production, utilizing various modern techniques such as molecular breeding, marker-assisted selection (MAS), GS, the role of genome editing (CRISPR-Cas), and transgenic approaches. Additionally, the chapter delves into prospects and significant challenges in the field of maize improvement.

2. Traditional approaches in maize crop improvement

Based on recent molecular analysis, it is now believed that the process of maize domestication commenced in the Central Balsas River Valley approximately 8700 long ago in southwestern Mexico. This domestication process occurred rapidly, originating from *Zea mays* ssp. *parviglumis* wild precursor, a subspecies of teosinte. This information is supported by studies conducted by Liu et al. [13], Piperno et al. [14], and Ranere et al. [15]. At the International Institute of Tropical Agriculture (IITA), maize underwent enhancements in various quantitative traits through classical or conventional methods. These improvements encompassed traits such as Striga resistance, nitrogen utilization efficiency, drought tolerance, resilience to stem borers, mitigation of aflatoxin accumulation, yield potential, and enhancement of nutritional quality [16, 17]. In traditional maize breeding, the approach entails the development of new plant cultivars by adhering to the principles of natural inheritance. This involves the selection of plants based on their exceptional performance in specific traits or characteristics, as discussed by Lamichhane and Thapa [18]. Conventional breeding methods have been employed for the two, self-pollinated and cross-pollinated plants for quite shortly. One example is the concept of pure line selection, which was introduced by Johannsen [19], as documented by Poehlman [20]. This method involves the creation of pure lines through the self-pollination of a single superior homozygous parental genotype. Following several years of conducting multi-locational trials, typically spanning approximately 6–7 years and involving the comparison with established check varieties, superior genotypes are officially introduced as new maize varieties. Pure-line selection is less effective due to low heritability caused by environmental effects, as genetic makeup closely resembles parental genotypes [21]. Mass selection, akin to pure-line selection, relies on highly heritable traits for plant choice [22]. Mass selection can be executed in two ways; the first is single-parental, where one kind of gamete is controlled, and the second one is bi-parental, where a couple of gametes, female and male are controlled. The chosen individuals are then planted in the crop land and harvested when they reach maturity. After harvest, seeds are mixed and sowed for the next generation. In the next year, crop plants grown from mixed seeds are justified with a check variety for variance. Selected plants are released as new varieties after multi-location trials. Backcross breeding introduces desirable traits from one plant into another without affecting other traits by crossing with a homozygous parent [23]. In this method, donor parents possess the desired trait, and recurrent parents receive these selected genes. After five to six generations of repeated backcrossing with the recurrent parent, the backcrossed progeny should inherit approximately 98% of the recurrent parent's genome [24]. In backcross breeding, the newly formed variety typically inherits a majority of its genes from the recurrent parent, with only a few coming from the donor parent, as noted by Singh [25]. Another method is recurrent selection, a term introduced by Hull [26], primarily applied in maize breeding but later extended to other cereal crops, as discussed by Ramya et al. [27]. This process entails the continued selection of favorable traits over multiple generations, to increase their prevalence through crosses between high-performing individuals from the heterozygous recurrent parent and inbred individuals, as discussed by Bangarwa [28]. Hybridization is another method for creating hybrids with desirable traits by mating genetically distant parents within the same species (Intraspecific hybridization) or between different species (Interspecific hybridization). It involves combining characteristics from different parents to produce genetically superior offspring, whether through natural or artificial means [21].

In a remarkable long-term study conducted with conventional breeding techniques, researchers at the Illinois Agricultural Experiment Station successfully enhanced the oil concentration in maize. They started with a base of approximately 5% oil content and, throughout 100 generations, developed high oil-producing maize lines, which now boast an impressive 20% oil content [29]. However, conventional breeding methods do have their limitations. For instance, identical parents do not produce variation due to the lack of segregation of gametes in conventional breeding [30]. Additionally, this process is often time-consuming, typically spanning over a decade or more before a new cultivar is ready for release, as noted by Bharti and Chimata [31]. Moreover, conventional breeding heavily relies on the cultivars phenotypic expressions to identify superior ones. Hence, the chosen cultivars may not consistently be without errors, given that phenotypes are significantly affected by genotype-environment interactions [32]. The selection process involves choosing individuals for breeding based on their differences in desired features, which are usually measurable or observable traits [33]. It's worth noting that conventional breeding is an applied science that heavily depends on the observations, skills, and experiences of breeders for judgment, as highlighted by Allard [34].

3. Molecular breeding and marker-assisted selection (MAS)

Traditional plant breeding involves the iterative practice of selecting both parents and their offspring based on desirable characteristics. The significance of molecular breeding is substantial in the contemporary world, especially in developing countries where a small proportion of the population is engaged in agriculture. This minority group bears the demanding responsibility of providing sustenance for the majority of the country's population [35]. This achievement is feasible because plant breeding has effectively enhanced crop yields without the need to expand the cultivated land or the workforce engaged in agriculture. Achieving this objective can be readily accomplished through crop enhancement via molecular breeding techniques. Molecular breeding employs various approaches such as the identification of simple traits or QTLs within breeding lines/populations, the integration of genes from breeding lines or wild relatives, gene pyramiding, marker-assisted recurrent selection (MARS), and Marker-assisted backcross (MABC).

Molecular MAS, often referred to as marker-aided selection or MAS is an indirect approach to nomination or selection wherein a specific trait is targeted through the use of a marker [29]. Within the context of MAS, a marker is located in the vicinity of a gene responsible for controlling the trait, thereby signifying the presence of a desirable allele when the marker is detected [36]. Knowing the alleles in key loci allows for the creation of optimal allele combinations to enhance the agronomic value of the genotype. Marker-assisted selection is commonly used for resistance gene pyramiding, which can provide complete resistance for several plant generations until it's challenged by pathogen strains. Achieving gene pyramiding for resistance becomes exceedingly challenging through classical breeding methods for certain traits like pathogen-induced disease resistance when dominant resistance genes are present [35]. There are primarily three categories of genetic markers. The first comprises visible or morphological markers, which are characteristics or phenotypic traits. The second category includes biochemical markers, which involve enzyme allelic variations referred to as isozymes. The third category consists of molecular or DNA markers, which unveil points of variation in the DNA sequence [37, 38]. For a nucleotide sequence to be useful as a molecular marker in molecular breeding, it usually requires polymorphism within its sequence. These variations in nucleotide sequences are unveiled through molecular methods like restriction fragment length polymorphism (RFLP), amplified fragment length polymorphism (AFLP), random amplified polymorphic DNA (RAPD), single nucleotide polymorphism (SNP), microsatellite or simple sequence repeat polymorphism (SSRP), sequence-tagged site (STS), single-strand conformation polymorphism (SSCP), and cleavable amplified polymorphic sequences (CAPS) among others [39]. Utilizing this marker set relies on extensive prior research. This typically involves various research stages for each trait, commencing with QTL mapping, progressing to fine mapping, and ultimately culminating in positional cloning [40]. For a successful MAS program, essential components include dependable markers, a robust DNA extraction method, well-constructed genetic maps, swift and efficient data handling, an interpretation of marker and trait connections, and knowledge to access tools for high-throughput marker detection [41].

MABC represents a specialized example of MAS, wherein the process of backcrossing is aided by molecular markers to expedite the selection of the recurrent parent and enhance genome recovery speed. The MABC technique has found extensive application in eliminating undesirable traits, such as susceptibility to insects and diseases, as well as anti-nutritional factors, from widely adopted high-yielding varieties by introducing QTLs or genes of interest from donor parents [42, 43]. Using DNA markers in a breeding program recommended a variety of benefits. For instance, DNA marker-based screening facilitates early selection for traits due to the evaluation of plant genotypes at the seedling stage or even from seeds, that may only manifest in adult plants, such as male sterility, fruit or grain quality, and photoperiod sensitivity. It expedites the selection of alleles that are difficult to assess phenotypically, especially for environmentally sensitive traits, simplifying and enhancing the breeding process. Individual plant selection, which may be impractical through phenotypic means, becomes feasible when relying on marker information. The issue of low heritability becomes inconsequential when using marker-based selection. Additionally, in traits with intricate inheritance patterns, it becomes possible to select each genetic component contributing to the trait independently. Using molecular markers, multiple characters that typically exhibit epistatic interactions can be preserved and ultimately stabilized. Moreover, the preservation of recessive genes does not necessitate progeny testing in every generation since homozygous and heterozygous plants can be differentiated using (codominant) markers, as explained by Lema [32]. Molecular markers have an important role in enhancing maize's genetic traits, such as addressing the intricate inheritance patterns related to drought tolerance [44–46], improving nutrient utilization [46–52], and enhancing diseases, parasitic and insect pests plant resistance in maize [17, 53–56]. Additional details can be found in the works of Hossain et al. [57] and Muntean et al. [33].

4. Genomic selection

The GS represents an advanced iteration of marker-assisted selection, enhancing the efficiency of selection and expediting the progress in selective breeding within a shorter timeframe. It achieves this by employing markers across the entire genome to predict the impact of quantitative gene loci, subsequently calculating genomic estimated breeding values (GEBVs) [58, 59]. The GS breeding approach was initially proposed by Meuwissen et al. [60]. GS can swiftly and cost-effectively forecast the yield potential of individual plants, ultimately leading to a reduction in both the time and expenses associated with a breeding cycle. In contrast to GWAS (genome-wide association studies) and linkage analysis, the primary goal of GS is not to pinpoint specific QTLs but rather to make predictions performance of offspring based on the DNA information gathered in the present. In GS, breeders can predict a plant's breeding value by utilizing data from all markers without the need for direct phenotype evaluation. This prediction relies on statistical models developed using a "training population" where both genotypes and phenotypes have been recorded. In mixed model analysis for genomic selection, markers are treated as random factors. This approach is necessary because the number of markers often exceeds the number of individuals in the training population, making it impractical to estimate the effect of each marker individually due to limited degrees of freedom [35, 61]. MAS proves valuable when choosing traits such as grain yields, flower colors, seed characteristics, and others that manifest primarily during the later reproductive stages. Through the application of GS, these traits can be detected by employing DNA markers in a genotype even at the preliminary stages of plant development as highlighted by Madhusudhana [62]. It offers numerous advantages compared to traditional MAS. Notably, it does not require QTL mapping, as it efficiently estimates breeding values using a comprehensive set of molecular markers that ideally spans the whole genome [63]; at early selection, it is more precise as it estimates all QTLs effects by utilizing high-density molecular markers and explains genetic variance for desirable traits. This stands in contrast to MAS, which relies on a limited number of markers for trait selection [64]; it shortens generation intervals, accelerates genetic progress (4–25% farther up phenotypic selection), and reduces breeding costs (26–56% lesser traditional methods) [65]; it exhibits superior efficiency in selecting traits with low heritability compared to MAS; in GS, breeding values serve as the selection criteria which are the sum of all allele genetic effects for each individual. This approach is known for its superior accuracy because it assesses the average performance of the offspring, rather than relying solely on the parents' performance [66]. In crops like maize, research conducted by the CIMMYT (International Maize and Wheat Improvement Center) [67], suggests that the breeding interval could be shortened to as little as half of the conventional timeframe. GS has been effectively applied to enhance various traits in maize, including shelling percentage, grain yield, grain moisture, ear length, ear width, ear rows, tassel branch number, kernels per spike, hundred-grain weight, kernel number per ear, and kernel depth [68, 69]. Challenges in GS arise from factors including the training population's size and variability, as well as the heritability of the traits being predicted. The statistical intricacies in GS are connected to the vast amount of marker data, where the number of markers far exceeds the number of observations [33]. In conclusion, GS has proven to be a valuable tool for the improvement of multiple important traits in maize, offering promising prospects for enhancing the overall performance and quality of maize crops.

5. Genome editing and CRISPR-Cas9

Genome editing is a form of genetic engineering that involves the intentional alteration of DNA within living cells through the insertion, deletion, or modification of genetic material. Genome editing, previously referred to as gene targeting, is a precise method for modifying the nucleotide sequence of the genome with an exceptional level of specificity down to the individual base pair. It encompasses a range of strategies and methods designed to make deliberate and customized modifications to the genetic makeup of an [70]. The essence of this editing technology is its reliance on site-specific nucleases (SSNs), which are customizable enzymes proficient in precisely targeting specific gene sequences. By utilizing these modified nucleases, it becomes feasible to precisely remove, insert, or replace particular gene sequences, demonstrating the benefits of site-directed mutagenesis when compared to random mutagenesis methods [71]. Genome editing techniques encompass a range of approaches, including tailored homing nucleases (meganucleases), zinc-finger nucleases (ZFNs), and transcription activator-like effector nucleases (TALENs). These methods employ protein-based systems that can be tailored to possess precise DNA-binding functions, allowing them to pinpoint specific gene sequences. The most widely adopted platform in recent times is clustered regularly interspaced short palindromic repeats-CRISPR-associated 9 nucleases (CRISPR-Cas9), which relies on RNA as a targeting element directing the nuclease to a specific DNA sequence [72, 73].

5.1 CRISPR-Cas9 in maize

Maize (*Zea mays* L.) is recognized as the most widely grown grain crop globally. Its versatility in terms of applications and adaptability to diverse environmental and soil conditions have contributed to its popularity as a desirable crop across the globe [74]. The drawbacks of random mutagenesis sparked investigations into precise genome modification methods, resulting in notable progress over the last decade. These methods have significantly increased the accuracy of gene editing, enhancing fidelity by nearly a thousandfold [75, 76]. The initial generation of targeted genome editing methods, including ZFNs and TALENs, achieved partial success but exhibited specific limitations [77].

In contrast, the CRISPR/Cas9 system, a relatively recent addition, has revolutionized the field of genome editing due to its simple design, operational flexibility, and cost-effectiveness [78–80]. The system consists of a universally used Cas9 nuclease protein and a solitary guide RNA (sgRNA) that includes a 20-base pair (bp) target site sequence along with a hairpin structure. The Cas9 protein induces a double-strand break (DSB) at the 20 bp genomic locus specified by the sgRNA, occurring near the NGG sequence called Protospacer Adjacent Motif (PAM), where N can represent any nucleotide. Cas9 consists of two catalytic nuclease domains, RuvC and HNH, responsible for creating DSBs at precise target sites guided by sgRNA. These DSBs can subsequently be repaired through two main mechanisms: Non-homologous end-joining (NHEJ) or Homology-directed repair (HDR) (see **Figure 1**). It's worth noting that a mere 20 bp sequence is ample for achieving allele specificity within single-copy regions of a genome, such as in maize. The compact size of the sequence simplifies the creation of sgRNA [81, 82].

In maize protoplasts, the CRISPR/Cas9 system achieved a targeted mutation efficiency of 13.1% in the phytic acid biosynthesis gene, ZmIPK, whereas TALENs achieved 9.1% [83]. Furthermore, the CRISPR technology demonstrated a mutation frequency in maize that was 10–20 times greater than that observed with homing endonucleases [84]. In a more recent advancement, a user-friendly public sector system known as "ISU Maize CRISPR" has been established to facilitate efficient site-specific mutagenesis in maize. It employs an *Escherichia coli* cloning vector and an Agrobacterium binary vector, enabling the incorporation of as many as four single guide RNAs (sgRNAs) for single or multiplex mutagenesis. This development marks

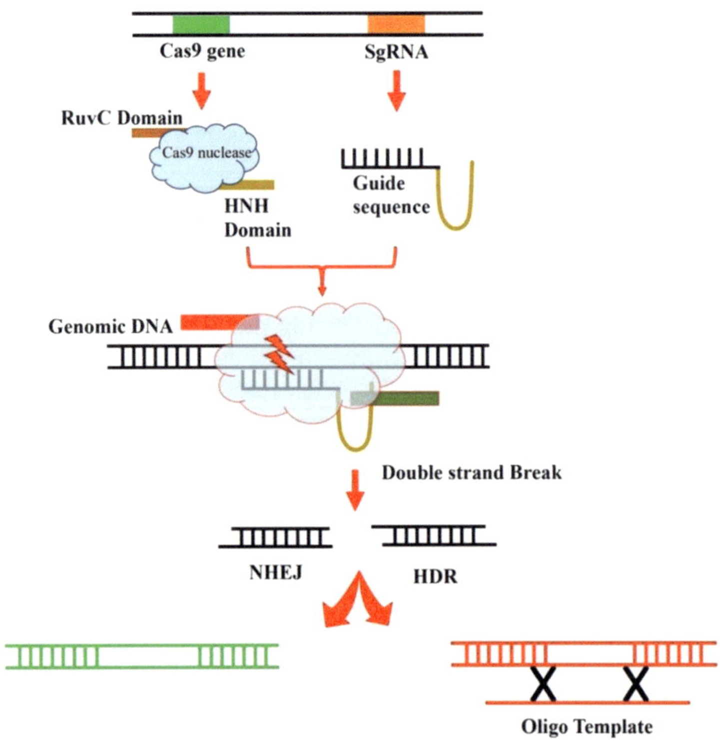

Figure 1.
A visual model depicting the genome modification process of the CRISPR/Cas9 system.

a significant stride in applying CRISPR/Cas9 for multifaceted gene editing in crops, with a specific focus on maize [85].

Delivering the CRISPR/Cas9 system, comprising the sgRNA and Cas9 protein, can be accomplished through transient techniques or by employing a long-term maize transformation process into the cell. Furthermore, this method allows for multiplexing [82]. Unlike some other nucleases, CRISPR/Cas9 is capable of targeting methylated DNA, rendering it a versatile tool for editing plant genomes [86].

5.2 Potential benefits of genome editing and ethical considerations

Genome editing starts from a molecular understanding of the target gene and utilizes a targeted and precise approach based on specific molecular knowledge. This allows genome editing to achieve highly accurate and intentional modifications in a controlled manner. One of the notable concerns linked to genome editing in plants is the potential for unintended genetic changes arising from off-target mutations [87, 88]. These off-target mutations refer to unintended changes in the DNA sequence of the plant's genome that occur at sites other than the intended target. Such unintended alterations can have unpredictable effects on the plant's characteristics and may raise safety and environmental concerns. The concerns surrounding genome editing also stem from the limited understanding of its principles and applications among the general public. There is a need for clear communication and education to address the knowledge gap and foster informed discussions about the technology. Differentiating between various categories of genetically modified plants, such as transgenic plants and genome-edited plants [89, 90], is essential. This distinction aids in comprehending the precise methods employed, the scope of genetic alterations made, and the possible consequences regarding safety, regulation, and public perception. By addressing these concerns and promoting transparency, scientists, policymakers, and stakeholders can work together to ensure responsible and ethically sound use of genome editing technologies in plant research and crop improvement. The risks related to the emerging techniques in genome editing encompass various interconnected aspects, including environmental, health-related, agricultural, economic, social, and political concerns. Among these, only a limited subset of risks is directly associated with the new techniques. One notable risk is the potential for bioterrorism, although it is currently only a theoretical concern when it comes to plants. Genome editing can have both positive and negative implications for agricultural risks, particularly in relation to biodiversity. On one hand, it has the potential to contribute to a reduction in biodiversity. On the other hand, it can also be utilized to enhance diversity and address emerging threats in agriculture.

6. Use of transgenic approaches to introduce foreign genes into maize

The adoption of genetically modified (GM) traits has been rapidly embraced by farmers worldwide, making it one of the fastest innovations in agriculture. As per findings presented by Brookes and Barfoot, [91], the worldwide economic benefits derived from genetically modified crop varieties amounted to US$ 225 billion during the period spanning from 1996 to 2018, with developing countries accounting for 52% of these gains. In 2019, transgenic crops were grown across 190.4 million hectares in 29 countries, intended for both food and feed purposes. This represents a substantial rise from 1.7 million hectares in 1996, marking a remarkable 112-fold increase.

The most widely adopted GM crops include soybean, maize, cotton, and canola. Of these crops, maize takes the lead. In 2019, 31% of the total global maize cultivation area, encompassing 60.9 million hectares across 14 countries, was allocated for growing genetically modified maize varieties [92].

In the last three decades, GM maize varieties have been effectively introduced to the market, providing farms with traits like resistance to herbicides and insects. The initial wave of genetically modified maize featured a single gene with a precise mechanism targeting a particular order of insects to confer insect resistance. In subsequent generations, the approach included crossbreeding herbicide and insect-resistant traits, as well as diverse insect-resistant traits, to establish multiple mechanisms of action against a range of insect orders. Farmers have found these stacked varieties to be remarkably effective, delivering evident and comprehensive phenotypic results [93]. Developing traits linked to quantitative characteristics such as tolerance to abiotic stress, efficient nutrient utilization, and increased yield presents a more intricate challenge. These characteristics are influenced by numerous genes and are sensitive to environmental factors, which adds complexity to the development process. To examine how individual genes affect complex traits, companies have established extensive biotechnology pipelines that involve evaluating genes in real field conditions on a large scale [94]. A standard biotechnology pipeline comprises various stages, which encompass discovery, demonstrating feasibility, initial development, advanced development, pre-launch, and the eventual release of commercial varieties. Certain stages might briefly coincide, particularly when a promising lead is identified early in the discovery phase, leading to the initiation of optimization efforts before validation is finished. The phase of gene discovery entails the quest for potential genes, which can be arduous, expensive, and uncertain, especially for traits such as drought tolerance and yield, which demand clearly defined phenotypic reactions. Extensive phenotypic screening of model plants like *Arabidopsis thaliana* and *Oryza sativa* is performed to assess hundreds of candidate genes [95, 96]. The proof-of-concept stage involves creating events for each candidate gene and conducting preliminary phenotypic evaluations in controlled settings as well as small-scale field experiments. During the early development phase, there is a focus on optimizing the lead to enhance stability and increase protein expression. Candidates demonstrating favorable agronomic performance, consistent trait expression, and heritability are chosen. These chosen candidates are subsequently subjected to molecular-level characterization and extensive field trials conducted across various locations and over multiple years [97].

In the advanced development stage, the validated leads are incorporated into commercial lines, often employing molecular markers to expedite the breeding process and guarantee the successful transfer of traits. In this phase, regulatory data related to the toxicity of gene products, allergenic potential, compositional analysis, as well as environmental and human safety aspects are collected. During the prelaunch stage, the production of seeds for the novel GM variety is expanded, quality control protocols are instituted to guarantee trait consistency and purity, a regulatory report is submitted, and arrangements are made for the commercial release of the new GM trait hybrid. The duration to finalize the pipeline, which varies according to the trait and available resources, generally averages around 11–13 years.

6.1 Enhancing traits in genetically modified (GM) maize varieties

Twenty-five years ago, the debut of the initial commercially accessible insect-resistant GM maize [98, 99] marked the beginning of a journey that led to the approval of 148 GM maize events for global commercial utilization [93]. By 2019, worldwide GM maize cultivation had expanded to cover 61 million hectares, with the most substantial acreage located in the USA (33 million hectares), Brazil (15 million hectares), Argentina (6 million hectares), and South Africa (2 million hectares) [92]. Among crops, maize holds the record for the largest number of approved GM events, totaling 148 events across 35 countries. Most of these events integrate traits like insect resistance and herbicide tolerance. Furthermore, approved traits for maize encompass fertility restoration, male sterility, heightened drought tolerance, phytase production, modified amino acids and alpha-amylase expression, improved photosynthesis, and increased ear biomass. These authorized traits encompass a total of 39 individual genes, with the largest proportion associated with insect resistance (18 genes) and herbicide tolerance (11 genes). The forthcoming generation of GM maize varieties poised for market release incorporates events featuring novel insecticidal proteins, including Vpb4Da2, DvSnf7 RNA, and IPDO72Aa. These proteins are designed to manage insect populations that have developed resistance to Bt [100–102]. Additional prospective varieties seek to enhance grain yield by upregulating the zmm28 and ZM-BG1H1 genes [103, 104], and to bolster drought tolerance through the overexpression of ARGOS8 [105].

6.2 Concerns and regulations related to GM crops

It's crucial to emphasize that scientific evidence confirms that GM crops do not present any heightened risks to both humans and the environment in comparison to conventional crops (National Academies of Sciences, Engineering, and Medicine, 2016). Nevertheless, public apprehensions and restrictions related to GM technology, notably within the European Union, continue to hold substantial importance [106]. To tackle these concerns and potentially shift public opinion, there is an ongoing exploration of new plant breeding technologies (NPBTs), including cisgenesis, intragenesis, and genome editing. Proficient communication of these technologies to the public holds the potential to impact public approval [107, 108]. Following 7 years of GM crop cultivation with no observable health impacts, apprehensions about potential environmental hazards, particularly gene transfer to other species, have gained more prominence than concerns regarding food safety. Pollen and seeds released into the environment may convey genetic characteristics to neighboring crops or wild relatives. Self-pollinating crops like wheat, barley, and potatoes have minimal chances of gene transfer, whereas cross-pollinating crops like sugar beets and corn are of greater concern in this regard. Although numerous cultivated crops lack wild counterparts in their present cultivation regions, the places of origin for these crop species are notably vulnerable to the infiltration of transgenic traits into native varieties or landraces. There is apprehension that transgenic varieties possessing a competitive edge might progressively supplant valuable genetic diversity. Consequently, Mexico, a nation harboring over 100 distinct corn varieties, has enforced a ban on the cultivation of transgenic corn.

7. Omics technologies in maize improvement

Advancements in the fields of biotechnology and computational sciences have paved the way for the generation of omics data on a large scale for various plant sets, including different varieties and species [109]. The application of diverse omics

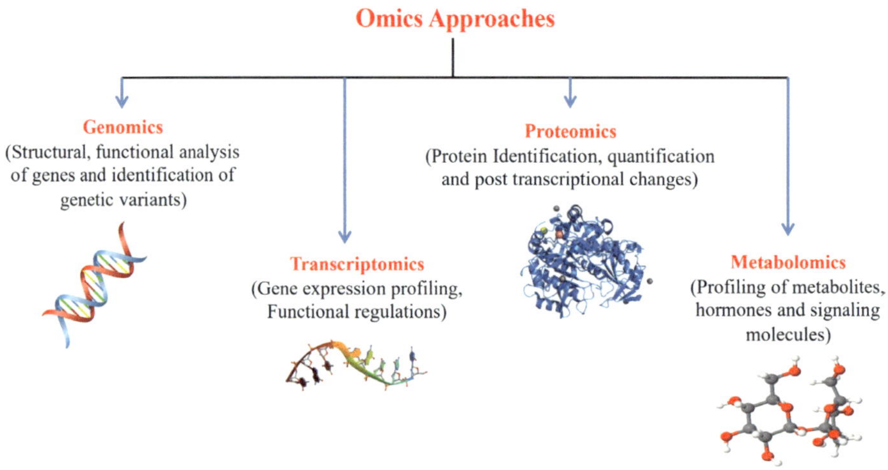

Figure 2.
Representation of different omics approaches used for maize crop improvement.

techniques has facilitated the discovery of genes, their respective functions, the specific types of RNA or proteins involved, their structural attributes, and the pathways influencing the development of final morphological traits. These identified genes can be subject to manipulation or transfer to create novel varieties or hybrids possessing advantageous traits. The multi-omics approach has proven successful in enhancing crop yields and developing resistance to stresses in agriculture (**Figure 2**). Molecular biology methods encompassing various omics technologies, such as genomics, transcriptomics, proteomics, and metabolomics, have played instrumental roles in advancing these research endeavors [110, 111].

Genomics is dedicated to the sequencing, characterization, and comprehensive exploration of a plant's genetic makeup, encompassing its composition, structure, functions, and intricate networks within the genome [112]. Novel approaches to plant breeding have been made possible by developments in plant genomics, which effectively enhance and expedite various aspects of the breeding process. These innovations include techniques such as marker-assisted selection, gene pyramiding, association mapping, breeding by design, genomic selection, and more [113–119]. In a study conducted by Vinayan and colleagues [120], genomic regions linked with fodder traits were pinpointed, and a prediction study on genomic regions was carried out using 1026 DH lines and 276 elite lines as prediction sets from bi-parental crosses.

In order to determine the expression profiles of both coding and non-coding RNA in response to different stresses, high-throughput sequencing platforms were used by transcriptomics to generate transcript data. It also incorporates RNA sequencing, microarray and serial analysis of gene expression (SAGE) [121]. The genetic makeup of the transcripts that have differential gene expression in particular cells has been revealed by a number of studies. These transcripts can affect phenotypic variations in maize, such as growth, yield components, disease tolerance, environmental response and quality traits. For instance, transcriptome correlation and comparisons signaling network analysis were used in a study by Liu and Zhang [122] to identify six genes essential to the control of the MAP Kinase cascade and HY5 module in the presence of blue light. In maize, these genes are essential for regulating stomata formation and dispersion. qRT-PCR and transcriptome analysis studies in maize roots, that are infected by *Holotrichia parallela* larvae, were the focus of another study project conducted in 2020 by Pan and colleagues [123, 124]. This showed the expression of twelve differently expressed genes linked to the pathways of benzoxazinoid production and Jasmonic acid-mediated signaling, which are in charge of maize roots' defense mechanisms against invaders. Zhou and colleagues [125] used bulked sergeant transcriptome analysis (BSTA) to study the mechanism behind maize's resistance to drought stress. On chromosome 2, four highly expressed candidate genes that confer *Gibberella* ear rot disease resistance in maize were found by transcriptome profiling of several inbred lines of maize [126]. Together, these results highlight the significance of transcriptomics in maize research, as it facilitates the discovery of essential regulatory components for enduring abiotic and biotic challenges, as well as the annotation of gene functions and the identification of candidate genes. Breeders will be able to solve present and future economic, ecological, and environmental concerns and ensure food security by using this information to gain the insights they need to create improved varieties of maize [69, 127, 128].

Proteomics involves a comprehensive investigation of proteins within a biological system, encompassing plants and animals, at a specific moment in time [129]. The analysis of proteomics serves the purpose of quantifying the abundance of various proteins, discerning alterations resulting from diverse post-translational modifications, and elucidating their functions and localization [130]. It offers a snapshot of diverse metabolic processes, their ensuing interactions, and their impacts on other regulatory pathways. Consequently, proteomic studies are indispensable for deciphering the diverse reactions within pathways under various stress conditions and time-frames [131]. The Proteomics field has attracted considerable interest from scientists seeking to examine physiological differences at the proteomic level under varying stress conditions. For example, Zhang et al. [132] performed a proteomic analysis of maize leaves in an attempt to evaluate proteome-level changes in corn when infected by the *Ostrinia furnacalis* (Asian corn borer). A total of 62 defense-responsive proteins were found, with a special focus on thioredoxin M-type and pathogenesis-related protein 1 (PR1), a chloroplastic precursor that significantly impacted the development of corn borer larvae and pupae. Comparative proteome profiling was done on resistant and susceptible lines exposed to *Puccinia polysora* (southern corn rust) in a study by Wang et al. [133]. This study demonstrated that resistance in the resistant lines was inhibited by a particular remorin protein (ZmREM 1.3). A comparative proteome profiling of drought-tolerant and susceptible maize lines was carried out by Dong et al. [134]. Plants use the development of defense-associated proteins (DAPs) in conjunction with the down-regulation of redundant proteins as a stress-reduction and energy-saving strategy.

Metabolomics is a cutting-edge biotechnique that seeks to identify functionally active metabolites, clarify their functions, and provide insight into the various biochemical processes that occur in plant genotypes and the phenotypic expressions that follow [135]. All metabolites, primary and secondary, with a molecular weight of less than 1500 Da, as well as their precursors and intermediates within the corresponding metabolic processes, are included in the metabolomes. Based on their particular goals, metabolic investigations can be divided into two categories: targeted and untargeted. The goal of targeted metabolomics is to precisely quantify one or a small number of metabolites from a predetermined list of recognized compounds. This method helps identify metabolites linked to particular features because of its high sensitivity and quantitative nature. Untargeted metabolomics, on the other hand, increases the possibility of identifying unintentional impacts by measuring the mass spectrometric properties of metabolites with unknown identities [136]. The application of metabolomics has proven invaluable in understanding how maize plants respond to various stress conditions i.e., heat, salinity and drought. For example, a metabolomic investigation involving salt-tolerant and salt-sensitive maize genotypes revealed differences in metabolite accumulation in both roots and seedlings under salt stress. In seedlings, salt stress induced glucose and acid metabolism. Thirty common chemicals, including metabolites linked to basic metabolism such as deoxyadenosine, adenine, L-pyroglutamic acid, cis-9-palmitoleic acid, and galactinol compounds, were found in the roots of both salt-sensitive and tolerant cultivars [137]. Heat stress effects on pollen male sterility in maize, especially at the most vulnerable tetrad stage, has also been clarified by metabolic pathway study. A reduction in pyruvate levels and an enhancement in sucrose levels were found in this research. In the meantime, other genes linked to signaling, unfolded protein stress, and auxin synthesis did not alter [138]. More importantly, a study by Ganie et al. [139] revealed metabolic pathways impacted by phosphorus stress situations, offering insights into strategies for improving phosphorus efficiency. The analysis, conducted using gas chromatography-mass spectroscopy, the investigation revealed a drop in fatty acids like cholesterol and stigmasterol, which are critical for membrane fluidity, and an increase in sugar alcohols like glucitol and mannitol under P-limitation, which are essential for membrane fluidity. In cases of severe phosphorus starvation, plants will scavenge phosphorus from these fatty acids, disrupting membrane fluidity. Additionally, elevated levels of serine and glycine indicated an increase in photorespiration rates.

8. Conclusion

The notable decrease in maize production resulting from various biotic and abiotic stresses raises substantial concerns. To address these challenges, farmers frequently turn to the application of chemical pesticides as they offer a rapid remedy. The negative consequences of widespread pesticide use on both human health and the environment have spurred the exploration of alternative pest control methods. Host plant resistance recognized as an eco-friendly strategy, has emerged as a vital component within Integrated Pest Management (IPM) initiatives. Maize varieties with stress resistance or tolerance offer a sustainable and environmentally conscious means of managing pests. Although strides have been made in pinpointing resistance sources for both biotic and abiotic stress in crops, traditional approaches for crafting such resilient or tolerant maize varieties are arduous and time-consuming. This is mainly attributable to the intricate character of quantitative traits, which are influenced by numerous genetic loci. Nevertheless, contemporary biotechnological tools, particularly various omics techniques, present encouraging prospects for creating sustainable, multi-faceted resistance to both biotic and abiotic stresses. Omics methodologies are currently being harnessed to develop novel plant resistance attributes that provide robust protection against various stresses in maize cultivation. This entails the utilization of innovative molecules, resistant genes, and the alteration of gene expression. The anticipated progress in biotechnological innovations, encompassing genome editing, genetic transformation, and marker-assisted breeding, among other methods, is poised to expedite the development of disease-resistant crops, both in the present and in the times ahead. RNA interference and genome editing through CRISPR/Cas9 represent novel approaches for creating disease-resistant crops. Biotechnology has emerged as a valuable instrument for tackling the worldwide pest challenge, resulting in the production of economically efficient, pesticide-resistant, and environmentally friendly insect-resistant crops. When employed judiciously and ethically, biotechnology holds the promise of delivering substantial advantages.

Conflict of interest

The authors declare no conflict of interest.

Author details

Vikas Kumar Singh[1], Saksham Pundir[1], Deepti Chaturvedi[1], Arshdeep Kaur[2], Abhishek Pandey[3], Shreya Mandal[4], Rahul Kumar[5], Ramesh Kumar Singh[6], Harish Pal Bhati[7], Parmdeep Singh Dhanda[8], Alpa Yadav[9], Chittaranjan Kole[10] and Prashant Kaushik[11]*

1 Department of Genetics and Plant Breeding, Ch. Charan Singh University, Meerut, Uttar Pradesh, India

2 Department of Botany, College of Basic Sciences and Humanities, Punjab Agricultural University, Ludhiana, Punjab, India

3 Department of Plant Breeding and Genetics, Punjab Agricultural University, Ludhiana, Punjab, India

4 Division of Biochemistry, ICAR-IARI, New Delhi, India

5 Department of Genetics and Plant Breeding, Sardar Vallabh Bhai Patel University of Agriculture and Technology, Meerut, Uttar Pradesh, India

6 Research and Development Centre, Eagle Seeds and Biotech Ltd., Indore, M.P., India

7 Department of Botany, Government College (Constituent), Jewar, G.B. Nagar, U.P., India

8 Department of Biochemistry, Punjab Agricultural University, Ludhiana, India

9 Department of Botany, Indira Gandhi University, Rewari, India

10 The Neotia University, Sarisha, India

11 Chaudhary Charan Singh Haryana Agricultural University, Hisar, India

*Address all correspondence to: prashantumri@gmail.com

References

[1] Erenstein O, Jaleta M, Sonder K, Mottaleb K, Prasanna BM. Global maize production, consumption and trade: Trends and R&D implications. Food Security. 2022;**14**:1295-1319

[2] Awika JM. Major cereal grains production and use around the world. In: Advances in Cereal Science: Implications to Food Processing and Health Promotion. Washington, DC: American Chemical Society; 2011. pp. 1-13

[3] FAOStat. FAO Stat. Rome: FAO; 2022. Available from: http://www.fao.org/faostat

[4] Grote U, Fasse A, Nguyen TT, Erenstein O. Food security and the dynamics of wheat and maize value chains in Africa and Asia. Frontiers in Sustainable Food Systems. 2021;**4**:617009

[5] Poole N, Donovan J, Erenstein O. Agri-nutrition research: Revisiting the contribution of maize and wheat to human nutrition and health. Food Policy. 2021;**100**:101976

[6] Serna-Saldivar SO, Perez-Carrillo E, Heredia-Olea E. Soybean-fortified wheat flour tortillas. In: Flour and Breads and their Fortification in Health and Disease Prevention. Cambridge, Massachusetts: Academic Press; 2019. pp. 291-306

[7] Tiwari A, Choudhary S, Padiya J, Ubale A, Mikkilineni V, Char B. Recent Advances and Applicability of GBS, GWAS, and GS in Maize. Genotyping by Sequencing for Crop Improvement. Hoboken, New Jersey: John Wiley & Sons; 2022. pp. 188-217

[8] Walbot V. Maize genome in motion. Genome Biology. UK: BioMed Central; 2008

[9] Schnable PS, Ware D, Fulton RS, Stein JC, Wei F, Pasternak S, et al. The B73 maize genome: Complexity, diversity, and dynamics. Science. 2009;**326**:1112-1115

[10] Jiao Y, Peluso P, Shi J, Liang T, Stitzer MC, Wang B, et al. Improved maize reference genome with single-molecule technologies. Nature. 2017;**546**:524-527

[11] Chapman MA, He Y, Zhou M. Beyond a reference genome: Pangenomes and population genomics of underutilized and orphan crops for future food and nutrition security. New Phytologist. 2022;**234**:1583-1597

[12] Voss-Fels KP, Cooper M, Hayes BJ. Accelerating crop genetic gains with genomic selection. Theoretical and Applied Genetics. 2019;**132**:669-686

[13] Liu C, Li X, Meng D, Zhong Y, Chen C, Dong X, et al. A 4-bp insertion at ZmPLA1 encoding a putative phospholipase A generates haploid induction in maize. Molecular Plant. 2017;**10**:520-522

[14] Piperno DR, Ranere AJ, Holst I, Iriarte J, Dickau R. Starch grain and phytolith evidence for early ninth millennium BP maize from the Central Balsas River Valley, Mexico. Proceedings of the National Academy of Sciences. 2009;**106**:5019-5024

[15] Ranere AJ, Piperno DR, Holst I, Dickau R, Iriarte J. The cultural and chronological context of early Holocene maize and squash domestication in the Central Balsas River Valley, Mexico. Proceedings of the National Academy of Sciences. 2009;**106**:5014-5018

[16] Alene AD, Menkir A, Ajala SO, Badu-Apraku B, Olanrewaju AS, Manyong VM, et al. The economic and poverty impacts of maize research in West and Central Africa. Agricultural Economics. 2009;**40**:535-550

[17] Gedil M, Menkir A. An integrated molecular and conventional breeding scheme for enhancing genetic gain in maize in Africa. Frontiers in Plant Science. 2019;**10**:1430

[18] Lamichhane S, Thapa S. Advances from conventional to modern plant breeding methodologies. Plant breeding and biotechnology. 2022;**10**:1-4

[19] Peters JA. Classic papers in genetics. Prentice-Hall, INC: Englewood Cliffs; 1959

[20] Poehlman JM. Breeding Field Crops. Berlin, Germany: Springer Science & Business Media; 17 Apr 2013

[21] Begna T. Conventional breeding methods widely used to improve self-pollinated crops. International Journal of Research. 2021;**7**:1-6

[22] Brown J, Caligari P. An Introduction to Plant Breeding. Hoboken, New Jersey: John Wiley & Sons; 2011

[23] Aleksoski J. The effect of backcross method in tobacco breeding. Journal of Agriculture and Plant Sciences. 2018;**16**:9-19

[24] Vogel KE. Backcross breeding. In: Transgenic Maize: Methods and Protocols. United State; 2009. pp. 161-169

[25] Singh SP. Alternative methods to backcross breeding. Annual Report of the Bean Improvement Cooperative. 1982;**25**:11-12

[26] Hull FH. Recurrent selection for specific combining ability in corn 1. Agronomy Journal. 1945;**37**:134-145

[27] Ramya P, Singh GP, Jain N, Singh PK, Pandey MK, Sharma K, et al. Effect of recurrent selection on drought tolerance and related morpho-physiological traits in bread wheat. PLoS One. 2016;**11**:e0156869

[28] Bangarwa S. Recurrent Selection - Definition and Types. North Carolina, USA; 2021. Available from: Biotecharticles.com

[29] Schlegel RH. History of Plant Breeding. United States: CRC Press; 2017

[30] Acquaah G. Principles of Plant Genetics and Breeding. Hoboken, New Jersey: John Wiley & Sons; 2009

[31] Bharti G, Chimata MK. Review on new plant breeding techniques. International Journal of Science and Research. 2019;**8**:723-730

[32] Lema M. Marker assisted selection in comparison to conventional plant breeding. Agricultural Research & Technology. 2018;**14**:555914

[33] Muntean L, Ona A, Berindean I, Racz I, Muntean S. Maize breeding: From domestication to genomic tools. Agronomy. 2022;**12**:2365

[34] Allard RW. Principles of plant breeding. Soil Science. 1961;**91**:414

[35] Breseghello F, Coelho AS. Traditional and modern plant breeding methods with examples in rice (*Oryza sativa* L.). Journal of Agricultural and Food Chemistry. 2013;**61**:8277-8286

[36] Stevens R. Prospects for using marker-assisted breeding to improve maize production in Africa. Journal of the Science of Food and Agriculture. 2008;**88**:745-755

[37] Jones CJ, Edwards KJ, Castaglione S, Winfield MO, Sala F, Van de Wiel C, et al.

Reproducibility testing of RAPD, AFLP and SSR markers in plants by a network of European laboratories. Molecular Breeding. 1997;**3**:381-390

[38] Winter P, Kahl G. Molecular marker technologies for plant improvement. World Journal of Microbiology and Biotechnology. 1995;**11**:438-448

[39] Mohan M, Nair S, Bhagwat A, Krishna TG, Yano M, Bhatia CR, et al. Genome mapping, molecular markers and marker-assisted selection in crop plants. Molecular Breeding. 1997;**3**:87-103

[40] Bhullar NK, Zhang Z, Wicker T, Keller B. Wheat gene bank accessions as a source of new alleles of the powdery mildew resistance gene Pm3: A large-scale allele mining project. BMC Plant Biology. 2010;**10**:1-3

[41] Kumawat G, Kumawat CK, Chandra K, Pandey S, Chand S, Mishra UN, et al. Insights into marker assisted selection and its applications in plant breeding. In: Plant Breeding-Current and Future Views. London, United Kingdom: IntechOpen; 2020

[42] Ribaut JM, Hoisington D. Marker-assisted selection: New tools and strategies. Trends in Plant Science. 1998;**3**:236-239

[43] Hasan MM, Rafii MY, Ismail MR, Mahmood M, Rahim HA, Alam MA, et al. Marker-assisted backcrossing: A useful method for rice improvement. Biotechnology & Biotechnological Equipment. 2015;**29**:237-254

[44] Ribaut JM, Ragot M. Marker-assisted selection to improve drought adaptation in maize: The backcross approach, perspectives, limitations, and alternatives. Journal of Experimental Botany. 2007;**58**:351-360

[45] Wang X, Wang H, Liu S, Ferjani A, Li J, Yan J, et al. Genetic variation in ZmVPP1 contributes to drought tolerance in maize seedlings. Nature Genetics. 2016;**48**:1233-1241

[46] Prasanna BM, Palacios-Rojas N, Hossain F, Muthusamy V, Menkir A, Dhliwayo T, et al. Molecular breeding for nutritionally enriched maize: Status and prospects. Frontiers in Genetics. 2020;**10**:1392

[47] Yathish KR, Karjagi CG, Gangoliya SS, Kumar A, Preeti J, Yadav HK, et al. Introgression of the low phytic acid locus (lpa2) into elite maize (Zea mays L.) inbreds through marker-assisted backcross breeding (MABB). Euphytica. Sep 2022;**218**(9):127

[48] Goswami R, Zunjare R, Khan S, Baveja A, Muthusamy V, Hossain F. Marker-assisted introgression of rare allele of crtRB1 gene into elite quality protein maize inbred for combining high lysine, tryptophan and provitamin A in maize. Plant Breeding. 2019;**138**:174-183

[49] Mehta BK, Muthusamy V, Zunjare RU, Baveja A, Chauhan HS, Chhabra R, et al. Biofortification of sweet corn hybrids for provitamin-A, lysine and tryptophan using molecular breeding. Journal of Cereal Science. 2020;**96**:103093

[50] Baveja A, Muthusamy V, Panda KK, Zunjare RU, Das AK, Chhabra R, et al. Development of multinutrient-rich biofortified sweet corn hybrids through genomics-assisted selection of *shrunken2, opaque2, lcyE* and *crtRB1* genes. Journal of Applied Genetics. 2021;**62**:419-429

[51] Das AK, Gowda MM, Muthusamy V, Zunjare RU, Chauhan HS, Baveja A, et al. Development of maize hybrids with enhanced vitamin-E, vitamin-A, lysine, and tryptophan through molecular

breeding. Frontiers in Plant Science. 2021;**12**:659381

[52] Singh J, Sharma S, Kaur A, Vikal Y, Cheema AK, Bains BK, et al. Marker-assisted pyramiding of lycopene-ε-cyclase, β-carotene hydroxylase1 and *opaque2* genes for development of biofortified maize hybrids. Scientific Reports. 2021;**11**:12642

[53] Zhao X, Tan G, Xing Y, Wei L, Chao Q, Zuo W, et al. Marker-assisted introgression of *qHSR1* to improve maize resistance to head smut. Molecular Breeding. 2012;**30**:1077-1088

[54] Li P. Understanding maize biology for better crop improvement. Molecular Plant. 2017;7:1014-1017

[55] Yang Q, He Y, Kabahuma M, Chaya T, Kelly A, Borrego E, et al. A gene encoding maize caffeoyl-CoA O-methyltransferase confers quantitative resistance to multiple pathogens. Nature Genetics. 2017;**9**:1364-1372

[56] Xu Z, Hua J, Wang F, Cheng Z, Meng Q, Chen Y, et al. Marker-assisted selection of *qMrdd8* to improve maize resistance to rough dwarf disease. Breeding Science. 2020;**70**:183-192

[57] Hossain F, Muthusamy V, Bhat JS, Zunjare RU, Kumar S, Prakash NR, et al. Maize breeding. In: Fundamentals of Field Crop Breeding. Singapore: Springer Nature Singapore; 2022. pp. 221-258

[58] Wang X, Xu Y, Hu Z, Xu C. Genomic selection methods for crop improvement: Current status and prospects. The Crop Journal. 2018;**6**:330-340

[59] Sun Q, Wang P, Li W, Li W, Lu S, Yu Y, et al. Genomic selection on shelling percentage and other traits for maize. Breeding Science. 2019;**69**(2):266-271

[60] Meuwissen TH, Hayes BJ, Goddard M. Prediction of total genetic value using genome-wide dense marker maps. Genetics. 2001;**157**:1819-1829

[61] Bernardo R, Yu J. Prospects for genome-wide selection for quantitative traits in maize. Crop Science. 2007;**3**:1082-1090

[62] Madhusudhana R. Marker-assisted breeding in sorghum. In: Breeding Sorghum for Diverse End Uses. Cambridge, United Kingdom; Woodhead Publishing; 2019. pp. 93-114

[63] Jannink JL, Lorenz AJ, Iwata H. Genomic selection in plant breeding: From theory to practice. Briefings in Functional Genomics. 2010;**9**:166-177

[64] Heffner EL, Sorrells ME, Jannink JL. Genomic selection for crop improvement. Crop Science. 2009;**49**:1-2

[65] Mayor PJ, Bernardo R. Genome wide selection and marker-assisted recurrent selection in doubled haploid versus F2 populations. Crop Science. 2009;**49**:1719-1725

[66] Massman JM, Jung HJ, Bernardo R. Genome-wide selection versus marker-assisted recurrent selection to improve grain yield and Stover-quality traits for cellulosic ethanol in maize. Crop Science. 2013;**53**:58-66

[67] Crossa J, Pérez-Rodríguez P, Cuevas J, Montesinos-López O, Jarquín D, De Los CG, et al. Genomic selection in plant breeding: Methods, models, and perspectives. Trends in Plant Science. 2017;**22**:961-975

[68] Wang T, Ma X, Li Y, Bai D, Liu C, Liu Z, et al. Changes in yield and yield components of single-cross maize

hybrids released in China between 1964 and 2001. Crop Science. 2011;**51**:512-525

[69] Sun G, Yu H, Wang P, Lopez-Guerrero M, Mural RV, Mizero ON, et al. A role for heritable transcriptomic variation in maize adaptation to temperate environments. Genome Biology. 2023;**24**:1-27

[70] Zaidi SK, Frietze SE, Gordon JA, Heath JL, Messier T, Hong D, et al. Bivalent epigenetic control of oncofetal gene expression in cancer. Molecular and Cellular Biology. 2017;**37**:e00352-e00317

[71] Osakabe K, Osakabe Y, Toki S. Site-directed mutagenesis in Arabidopsis using custom-designed zinc finger nucleases. Proceedings of the National Academy of Sciences. 2010;**107**:12034-12039

[72] Voytas DF, Gao C. Precision genome engineering and agriculture: Opportunities and regulatory challenges. PLoS Biology. 2014;**12**:e1001877

[73] Sander JD, Joung JK. CRISPR-Cas systems for editing, regulating and targeting genomes. Nature Biotechnology. 2014;**32**:347-355

[74] Pegoraro C, Mertz LM, da Maia LC, Rombaldi CV, de Oliveira AC. Importance of heat shock proteins in maize. Journal of Crop Science and Biotechnology. 2011;**14**:85-95. DOI: 10.1007/s12892-010-0119-3

[75] Puchta H, Dujon B, Hohn B. Homologous recombination in plant cells is enhanced by in vivo induction of double strand breaks into DNA by a site-specific endonuclease. Nucleic Acids Research. 1993;**21**:5034-5040. DOI: 10.1093/nar/21.22.5034

[76] Svitashev S, Schwartz C, Lenderts B, Young JK, Cigan AM. Genome editing in maize directed by CRISPR–Cas9 ribonucleoprotein complexes. Nature Communications. 2016;**7**:13274. DOI: 10.1038/ncomms13274

[77] Joung JK, Sander JD. TALENs: A widely applicable technology for targeted genome editing. Nature Reviews. Molecular Cell Biology. 2013;**14**:49-55. DOI: 10.1038/nrm3486

[78] Jinek M, Chylinski K, Fonfara I, Hauer M, Doudna JA, Charpentier E. A programmable dual-RNA–guided DNA endonuclease in adaptive bacterial immunity. Science. 2012;**337**:816-821. DOI: 10.1126/science.1225829

[79] Gilbert LA, Larson MH, Morsut L, Liu Z, Brar GA, Torres SE, et al. CRISPR-mediated modular RNA-guided regulation of transcription in eukaryotes. Cell. 2013;**154**:442-451. DOI: 10.1016/j.cell.2013.06.044

[80] O'Connell MR, Oakes BL, Sternberg SH, East-Seletsky A, Kaplan M, Doudna JA. Programmable RNA recognition and cleavage by CRISPR/Cas9. Nature. 2014;**516**:263-266. DOI: 10.1038/nature13769

[81] Newman M, Ausubel FM. Introduction to gene editing and manipulation using CRISPR/Cas9 technology. Current Protocols in Molecular Biology. 2016;**115**:1-6

[82] Shan Q, Wang Y, Li J, Gao C. Genome editing in rice and wheat using the CRISPR/Cas system. Nature Protocols. 2014;**9**:2395-2410. DOI: 10.1038/nprot.2014.157

[83] Liang Z, Zhang K, Chen K, Gao C. Targeted mutagenesis in *Zea mays* using TALENs and the CRISPR/Cas system. Journal of Genetics and Genomics. 2014;**41**:63-68. DOI: 10.1016/j.jgg.2013.12.001

[84] Svitashev S, Young JK, Schwartz C, Gao H, Falco SC, Cigan AM. Targeted mutagenesis, precise gene editing, and site-specific gene insertion in maize using Cas9 and guide RNA. Plant Physiology. 2015;**169**:931-945. DOI: 10.1104/pp.15.00793

[85] Char SN, Neelakandan AK, Nahampun H, Frame B, Main M, Spalding MH, et al. An Agrobacterium-delivered CRISPR/Cas9 system for high-frequency targeted mutagenesis in maize. Plant Biotechnology Journal. 2017;**15**:257-268. DOI: 10.1111/pbi.12611

[86] Bortesi L, Fischer R. The CRISPR/Cas9 system for plant genome editing and beyond. Biotechnology Advances. 2015;**33**:41-52. DOI: 10.1016/j.biotechadv.2014.12.006

[87] Liang Z, Chen K, Zhang Y, Liu J, Yin K, Qiu JL. Genome editing of bread wheat using biolistic delivery of CRISPR/Cas9 in vitro transcripts or ribonucleoproteins. Nature Protocols. 2018;**13**:413-430. DOI: 10.1038/nprot.2017.145

[88] Pineda M, Lear A, Collins JP, Kiani S. Safe CRISPR: Challenges and possible solutions. Trends in Biotechnology. 2019;**37**:389-401. DOI: 10.1016/j.tibtech.2018.09.010

[89] Garcia RMT, Knapp AN, Garcia-Ruiz H. Profile of genetically modified plants authorized in Mexico. GM Crops Food. 2018;**9**:152-168. DOI: 10.1080/21645698.2018.1507601

[90] Eckerstorfer MF, Engelhard M, Heissenberger A, Simon S, Teichmann H. Plants developed by new genetic modification techniques comparison of existing regulatory frameworks in the EU and non-EU countries. Frontiers in Bioengineering and Biotechnology. 2019;7:26. DOI: 10.3389/fbioe.2019.00026

[91] Brookes G, Barfoot P. GM Crops: Global Socio-Economic and Environmental Impacts. United Kingdom: Taylor & Francis; 2020. pp. 1996-2018

[92] ISAAA Brief. Global Status of Commercialized Biotech/GM Crops in 2019. NY: ISAAA Ithaca; 2019

[93] ISAAA Database. GM Approval Database. 2021. Available from: https://www.isaaa.org/gmapprovaldatabase/default.asp

[94] Simmons CR, Lafitte HR, Reimann KS, Brugière N, Roesler K, Albertsen MC. Successes and insights of an industry biotech program to enhance maize agronomic traits. Plant Science. 2021;**307**:110899. DOI: 10.1016/j.plantsci.2021.110899

[95] Wang A, Shu X, Jing X, Jiao C, Chen L, Zhang J, et al. Identification of rice (*Oryza sativa* L.) genes involved in sheath blight resistance via a genome-wide association study. Plant Biotechnology Journal. 2021;**19**:1553-1566

[96] Kumar V, Singh A, Mithra SA, Krishnamurthy SL, Parida SK, Jain S, et al. Genome-wide association mapping of salinity tolerance in rice (*Oryza sativa*). DNA Research. 2015;**22**:133-145

[97] Alam MA, Rahman M, Ahmed S, Jahan N, Khan MA, Islam MR, et al. Genetic variation and genotype by environment interaction for agronomic traits in maize (*Zea mays* L.) hybrids. Plants. 2022;**11**:1522

[98] Tabashnik BE, Carrière Y. Surge in insect resistance to transgenic crops and prospects for sustainability. Nature Biotechnology. 2019;**35**:926-935. DOI: 10.1038/nbt.3974

[99] Pellegrino E, Bedini S, Nuti M, Ercoli L. Impact of genetically

engineered maize on agronomic, environmental and toxicological traits: A metaanalysis of 21 years of field data. Scientific Reports. 2018;**8**:3113. DOI: 10.1038/s41598-018-21284-2

[100] Schellenberger U, Oral J, Rosen BA, Wei JZ, Zhu G, Xie W. A selective insecticidal protein from pseudomonas for controlling corn rootworms. Science. 2016;**354**:634-637. DOI: 10.1126/science.aaf6056

[101] Moar W, Khajuria C, Pleau M, Ilagan O, Chen M, Jiang C. Cry3Bb1-resistant western corn rootworm, *diabrotica virgifera virgifera* (LeConte) does not exhibit cross-resistance to DvSnf7 dsRNA. PLoS One. 2018;**12**:1-15. DOI: 10.1371/journal.pone.0169175

[102] Yin Y, Flasinski S, Moar W, Bowen D, Chay C, Milligan J. A new bacillus thuringiensis protein for Western corn rootworm control. PLoS One. 2020;**15**:1-16. DOI: 10.1371/journal.pone.0242791

[103] Wu J, Lawit SJ, Weers B, Sun J, Mongar N, Van Hemert J. Overexpression of zmm28 increases maize grain yield in the field. Proceedings of the National Academy of Sciences. 2019;**116**:23850-23858. DOI: 10.1073/pnas.1902593116

[104] Simmons CR, Weers BP, Reimann KS, Abbitt SE, Frank MJ, Wang W. Maize BIG GRAIN1 homolog overexpression increases maize GRAIN yield. Plant Biotechnology Journal. 2020;**18**:2304-2315. DOI: 10.1111/pbi.13392

[105] Shi J, Habben JE, Archibald RL, Drummond BJ, Chamberlin MA, Williams RW. Overexpression of ARGOS genes modifies plant sensitivity to ethylene, leading to improved drought tolerance in both Arabidopsis and maize. Plant Physiology. 2015;**169**:266-282. DOI: 10.1104/pp.15.00780

[106] Woźniak E, Tyczewska A, Twardowski T. A shift towards biotechnology: Social opinion in the EU. Trends in Biotechnology. 2021;**39**:214-218. DOI: 10.1016/j.tibtech.2020.08.001

[107] Cardi T. Cisgenesis and genome editing: Combining concepts and efforts for a smarter use of genetic resources in crop breeding. Plant Breeding. 2016;**135**:139-147. DOI: 10.1111/pbr.12345

[108] Harfouche AL, Petousi V, Meilan R, Sweet J, Twardowski T, Altman A. Promoting ethically responsible use of agricultural biotechnology. Trends in Plant Science. 2021;**26**:546-559. DOI: 10.1016/j.tplants.2020.12.015

[109] Li Q, Yan J. Sustainable agriculture in the era of omics: Knowledge-driven crop breeding. Genome Biology. 2020;**21**:154

[110] Somegowda VK, Rayaprolu L, Rathore A, Deshpande SP, Gupta R. Genome-wide association studies (GWAS) for traits related to fodder quality and biofuel in sorghum: Progress and prospects. Protein and Peptide Letters. 2021;**28**:843-854

[111] Al-Qurainy F, Alshameri A, Gaafar AR, Khan S, Nadeem M, Alameri AA, et al. Comprehensive stress-based De novo transcriptome assembly and annotation of guar (*Cyamopsis tetragonoloba* L. Taub.): An important industrial and forage crop. International Journal of Genomics. 2019;**2019**:7295859

[112] Abdurakhmonov IY. Plant Genomics. London, UK: IntechOpen; 2016. ISBN 978-953-51-2455-9

[113] Pérez-de-Castro AM, Vilanova S, Cañizares J, Pascual L, Blanca JM, Díez MJ, et al. Application of genomic tools in plant breeding. Current Genomics. 2012;**13**:179-195

[114] Singh VK, Chaturvedi D, Pundir S, Kumar D, Sharma R, Kumar S, et al. GWAS scans of cereal cyst nematode (*Heterodera avenae*) resistance in Indian wheat germplasm. Molecular Genetics and Genomics. 2023;**298**:579-601

[115] Chaturvedi D, Pundir S, Singh VK, Kumar D, Sharma R, Röder MS, et al. Identification of genomic regions associated with cereal cyst nematode (*Heterodera avenae* Woll.) resistance in spring and winter wheat. Scientific Reports. 2023;**13**:5916

[116] Kaur H, Sharma P, Kumar J, Singh VK, Vasistha NK, Gahlaut V, et al. Genetic analysis of iron, zinc and grain yield in wheat-Aegilops derivatives using multi-locus GWAS. Molecular Biology Reports. 2023;**50**:1-2

[117] Kumar D, Sharma S, Sharma R, Pundir S, Singh VK, Chaturvedi D, et al. Genome-wide association study in hexaploid wheat identifies novel genomic regions associated with resistance to root lesion nematode (*Pratylenchus thornei*). Scientific Reports. 2021;**11**:3572

[118] Pundir S, Sharma R, Kumar D, Singh VK, Chaturvedi D, Kanwar RS, et al. QTL mapping for resistance against cereal cyst nematode (*Heterodera avenae* Woll.) in wheat (*Triticum aestivum* L.). Scientific Reports. 2022;**12**:9586

[119] Pundir S, Singh R, Singh VK, Sharma S, Balyan HS, Gupta PK, et al. Mapping of QTLs and meta-QTLs for *Heterodera avenae* Woll. resistance in common wheat (*Triticum aestivum* L.). BMC Plant Biology. 2023;**23**:529

[120] Vinayan MT, Seetharam K, Babu R, Zaidi PH, Blummel M, Nair SK. Genome wide association study and genomic prediction for Stover quality traits in tropical maize (*Zea mays* L.). Scientific Reports. 2021;**11**:686

[121] Cramer GR, Urano K, Delrot S, Pezzotti M, Shinozaki K. Effects of abiotic stress on plants: A systems biology perspective. BMC Plant Biology. 2011;**11**:163

[122] Liu T, Zhang X. Transcriptome and metabolomic analyses reveal regulatory networks controlling maize stomatal development in response to blue light. International Journal of Molecular Sciences. 2021;**22**:5393

[123] Pan Y, Zhao SW, Tang XL, Wang S, Wang X, Zhang XX, et al. Transcriptome analysis of maize reveals potential key genes involved in the response to belowground herbivore *Holotrichia parallela* larvae feeding. Genome. 2020;**63**:1-2

[124] Pan Y, Zhao S, Wang Z, Wang X, Zhang X, Lee Y, et al. Quantitative proteomics suggests changes in the carbohydrate metabolism of maize in response to larvae of the belowground herbivore *Holotrichia parallela*. Peer J. 2020;**8**:e9819

[125] Zhou K, Zeng X, Zhang B, Aslam M, Xin H, Liu W, et al. Bulk segregant transcriptome analysis based differential expression of drought response genes in maize. Pakistan Journal of Agricultural Sciences. 2020;**57**:909-923

[126] Kebede AZ, Johnston A, Schneiderman D, Bosnich W, Harris LJ. Transcriptome profiling of two maize inbreds with distinct responses to *Gibberella* ear rot disease to identify candidate resistance genes. BMC Genomics. 2018;**19**:131

[127] He W, Zhu Y, Leng Y, Yang L, Zhang B, Yang J, et al. Transcriptomic analysis reveals candidate genes responding maize gray leaf spot caused by *Cercospora zeina*. Plants. 2021;**10**:2257

[128] Wu X, Wang B, Xie F, Zhang L, Gong J, Zhu W, et al. QTL mapping and transcriptome analysis identify candidate genes regulating pericarp thickness in sweet corn. BMC Plant Biology. 2020;**20**:117

[129] Suhre K, McCarthy MI, Schwenk JM. Genetics meets proteomics: Perspectives for large population-based studies. Nature Reviews. Genetics. 2021;**22**:19-37

[130] Mustafa G, Komatsu S. Plant proteomic research for improvement of food crops under stresses: A review. Molecular Omics. 2021;**17**:860-880

[131] Guo J, Li C, Zhang X, Li Y, Zhang D, Shi Y, et al. Transcriptome and GWAS analyses reveal candidate gene for seminal root length of maize seedlings under drought stress. Plant Science. 2020;**292**:110380

[132] Zhang YT, Zhang YL, Chen SX, Yin GH, Yang ZZ, Lee S, et al. Proteomics of methyl jasmonate induced defense response in maize leaves against Asian corn borer. BMC Genomics. 2015;**16**:224

[133] Wang J, Lin Z, Zhang X, Liu H, Zhou L, Zhong S, et al. krn1, a major quantitative trait locus for kernel row number in maize. New Phytologist. 2019;**223**:1634-1646

[134] Dong A, Yang Y, Liu S, Zenda T, Liu X, Wang Y, et al. Comparative proteomics analysis of two maize hybrids revealed drought-stress tolerance mechanisms. Biotechnology & Biotechnological Equipment. 2020;**34**:763-780

[135] Führs H, Götze S, Specht A, Erban A, Gallien S, Heintz D, et al. Characterization of leaf apoplastic peroxidases and metabolites in *Vigna unguiculata* in response to toxic manganese supply and silicon. Journal of Experimental Botany. 2009;**60**:1663-1678

[136] Kaur B, Sandhu KS, Kamal R, Kaur K, Singh J, Röder MS, et al. Omics for the improvement of abiotic, biotic, and agronomic traits in major cereal crops: Applications, challenges, and prospects. Plants. 2021;**10**:1989

[137] Yue JY, Wang LH, Dou XT, Wang YJ, Wang HZ. Comparative metabolomic profiling in the roots of salt-tolerant and salt-intolerant maize cultivars treated with NaCl stress. Biologia Plantarum. 2020;**64**:569-577

[138] Begcy K, Nosenko T, Zhou LZ, Fragner L, Weckwerth W, Dresselhaus T. Male sterility in maize after transient heat stress during the tetrad stage of pollen development. Plant Physiology. 2019;**181**:683-700

[139] Ganie AH, Ahmad A, Pandey R, Aref IM, Yousuf PY, Ahmad S, et al. Metabolite profiling of low-P tolerant and low-P sensitive maize genotypes under phosphorus starvation and restoration conditions. PLoS One. 2015;**10**:e0129520

Chapter 8

Corn for Biofuel: Status, Prospects and Implications

Smruti Ranjan Padhan, Shankar Lal Jat, Pratikshya Mishra, Sibananda Darjee, Sushmita Saini, Soumya Ranjan Padhan, Radheshyam and Shivani Ranjan

Abstract

Biofuel offers an alternative energy source to meet the energy demands of a growing population of 8 billion while minimizing environmental impact. Globally, around 3000 petajoules of biofuel are produced, diversifying energy sources from conventional to renewable. Corn, rich in starch that can be converted into ethanol, is widely used in biofuel production. Corn-based biofuels are popular due to their potential to reduce greenhouse gas emissions, their biodegradability, and clean ignition, enhancing energy security. While the current state of corn as a biofuel source appears promising, increasing production requires breeding strategies like varietal crossing and cultivar selection to enhance biomass and starch content. Better agronomic practices and extension strategies are also necessary to improve yield and promote adoption among farmers. Using maize as a feedstock for biofuel production can boost the agricultural industry, create jobs in farming, processing, and transportation, and reduce reliance on foreign oil while preserving foreign exchange reserves. Technological advancements, viz., cellulosic ethanol production, have further expanded the potential use of corn for biofuels due to its abundance and convenience. However, the future of corn-based biofuels is uncertain. Therefore, ongoing innovation, exploration of alternative feedstocks, and cutting-edge technologies are necessary to overcome challenges.

Keywords: breeding approaches, climate change mitigation, energy security, ethanol, food security, starch

1. Introduction

Maize, also known as corn, since its domestication from 9000 years ago, is one of the most important cereal crops globally, serving as a staple food for millions of people and a valuable feedstock for livestock [1]. Corn (dry grain) is annually cultivated on a projected 206 Mha of land worldwide, making it the second most extensively grown crop globally after wheat with a peak production and productivity of 1210 million tonnes and 5879 kg/ha, respectively [2]. Taking into consideration of area stagnation of wheat and rice, maize is set to overtake wheat in terms of acreage by 2030 [3]. Notably, the Americas, encompassing the United States, Brazil, and Argentina, are significant maize producers due to their favorable climate, extensive agricultural infrastructure, and technological advancements along with other important maize-producing regions involving China, India and South Africa, where suitable agro-climatic conditions and dedicated cultivation efforts contribute to substantial maize yields [4]. Corn production plays a diverse and critical role in global food security, economic development, and agricultural sustainability. It is widespread across the globe, with specific regions emerging as major contributors to the global supply. Trend analysis has shown that the maize production has increasing significantly thanks to upsurge in productivity (by six times from 1961 to 2021) making it a potential crop for alternative use like biofuel (**Figure 1**).

Since being necessary is the essence of the invention, the impending energy crisis has sparked curiosity about the production of biofuel. In the upcoming years, there will be a dramatic increase in the global consumption of liquid petroleum. By 2025, the energy demand is predicted to increase by more than 50% if the current trend holds [5]. Most crucially, an endless need for finite petroleum resources cannot be a long-term viable solution. Therefore, we must begin working towards carrying a switch from the non-renewable energy source of carbon to renewable vis-à-vis sustainable bio-resources prior to situations starting to slip out of our purview. In this way, the corn-to-biofuel notion can serve as a road map.

Biofuel, derived from renewable biomass sources, has emerged as an important alternative to fossil fuels by diversifying the energy sources [6] due to its potential to mitigate climate change, enhance energy security, and promote sustainable development [7]. Because vegetation assimilates CO_2 over its growth phase during the reaction of photosynthesis biofuels have been recognized as carbon-neutral energy sources [8]. In this context, biofuels, a healthier option to petroleum, are gaining popularity because they are environmentally friendly and compassionate to the environment. Utilizing these fuels may assist reduce environmental changes and the pollutants that come from automobiles. Biofuels are produced through various methods, including biomass conversion processes such as fermentation, pyrolysis, and transesterification. Biomass feedstocks can range from crops like sugarcane,

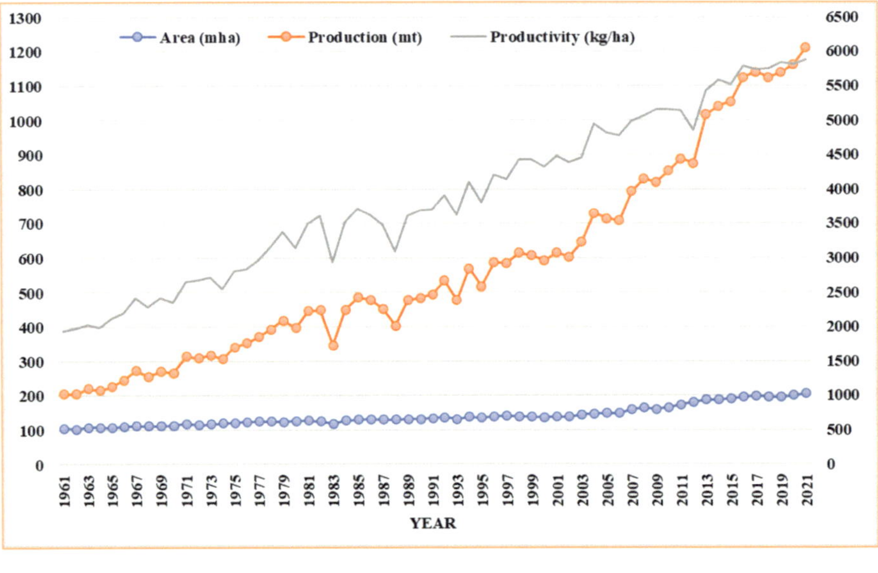

Figure 1.
Area, production, productivity trend of corn from 1961 to 2021 (data collected from http://www.fao.org/faostat).

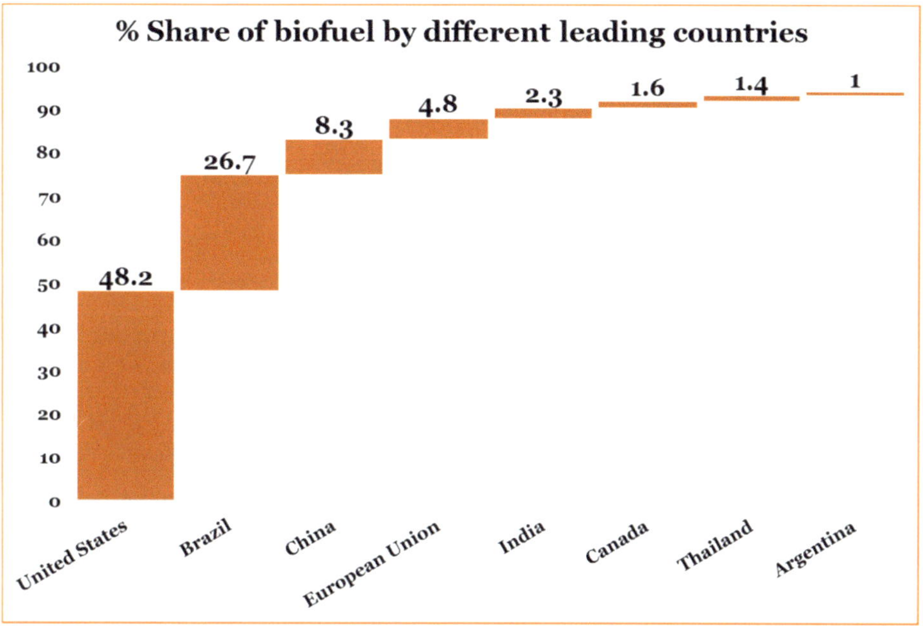

Figure 2.
Percentage share of biofuel by different leading countries (data obtained from OECD-FAO, 2021 [13]).

corn, and soybeans, to agricultural residues, forest residues, and even algae [9]. These feedstocks are transformed into liquid fuels, such as ethanol and biodiesel, or gaseous fuels like biogas. The two most prevalent kinds of biofuels employed nowadays are ethanol and biodiesel, which fall under the first generation of biofuels. Other types of biofuels include methanol, biodiesel, biogas, and Syngas. Ethanol is produced from cellulose-based feedstocks [10] such as corn, sugarcane, discarded potatoes, and others [11]. It is frequently used with petrol as an additional ingredient to raise the octane level and lower greenhouse gas emissions (GHG) [12]. With outputs totaling 1436 petajoules in 2021, the United States was the world's top manufacturer of biofuel followed by Brazil with statistics of about 840 petajoules, sharing 48.2 and 26.7%, respectively of the world total biofuel production (**Figure 2**, [13]).

Corn has been a major crop in the United States and many other countries around the world. In recent years, it has also become an important source of biofuel production due to its high starch content that can be converted into ethanol through different processes by fermentation [14]. The use of corn for biofuels has gained significant attention as an alternative to fossil fuels, which has the potential to reduce GHG emissions and promote energy security. However, the use of corn for biofuels has also raised concerns regarding its impact on food security, the environment, and the economy.

2. Corn biofuel: current status

2.1 Historical overview of corn for biofuels

The history of corn for biofuels is a long and complex one with its use dating back centuries, but it was only in the twentieth century that it gained significant attention as an alternative to fossil fuels. In the early 1900s, Henry Ford experimented with ethanol as a fuel for his Model T, and during World War II, the US government promoted the use of ethanol as a way to address gasoline shortages. However, the high cost of producing ethanol from corn made it economically unfeasible, and the program was discontinued after the war. Further, in the 1970s, the oil crisis sparked renewed interest in biofuels, and corn ethanol once again became a topic of discussion [14, 15]. In 1978, the US Congress passed the Energy Tax Act, which provided tax incentives for ethanol production and mandated the use of ethanol in gasoline. However, the impact of the act was limited, and ethanol production remained relatively low.

The twenty first century saw a significant increase in the use of corn for biofuels, driven by concerns over climate change and energy security. In 2005, the US government enacted the Renewable Fuel Standard, which mandated the use of biofuels, including corn ethanol, in transportation fuels [16]. The mandate has since been expanded, with the goal of reaching 36 billion gallons of biofuels by 2022. The use of corn for biofuels has had a significant impact on global markets, particularly in the US, which is the largest producer of corn ethanol [12]. The increased demand for corn as a feedstock has led to higher corn prices and raised concerns about food security. While corn ethanol has the potential to reduce (GHG) emissions and promote energy security, its production has been criticized for its potential negative impact on food security, environmental impacts, and global markets.

2.2 Current trends in corn-based biofuel production

The corn biofuel production process is complex and requires careful management to ensure that the final product is of high quality and meets regulatory requirements (**Figure 3**). With advances in technology and process optimization, the production process can be made more efficient and environmentally sustainable.

Corn-based biofuel production is a dynamic and constantly evolving industry and its production has been a budding industry over the past decade, with the United States leading the way as the world's largest producer of ethanol from corn. The increasing demand for renewable fuel sources has led to a significant increase in corn-based biofuel production, with corn ethanol being the most widely used form of biofuel in the US. One of the major trends in corn-based biofuel production is the development of new technologies that allow for more efficient and cost-effective production. This includes the use of genetically modified corn that has been specifically bred to produce higher yields of ethanol, as well as the use of more efficient production processes that reduce energy and water usage [17, 18].

The development of new markets and applications for the fuel has set new vistas for corn-based biofuel production. In addition to its use as a transportation fuel, corn ethanol is being used in a variety of industrial applications, such as solvents, cleaning agents, and as a feedstock for the production of other chemicals [19]. There is also a growing inclination towards the use of advanced biofuels, which are produced from non-food sources such as switchgrass and algae with having the potential to be more sustainable and environmentally and are turning out to be the better competitors of corn-based biofuels [20]. Despite these challenges, the corn-based biofuel industry remains a vital component of the global energy mix. With continued investment in research and development, the future of corn-biofuel production looks promising, as long as the industry continues to prioritize sustainability and responsible production practices.

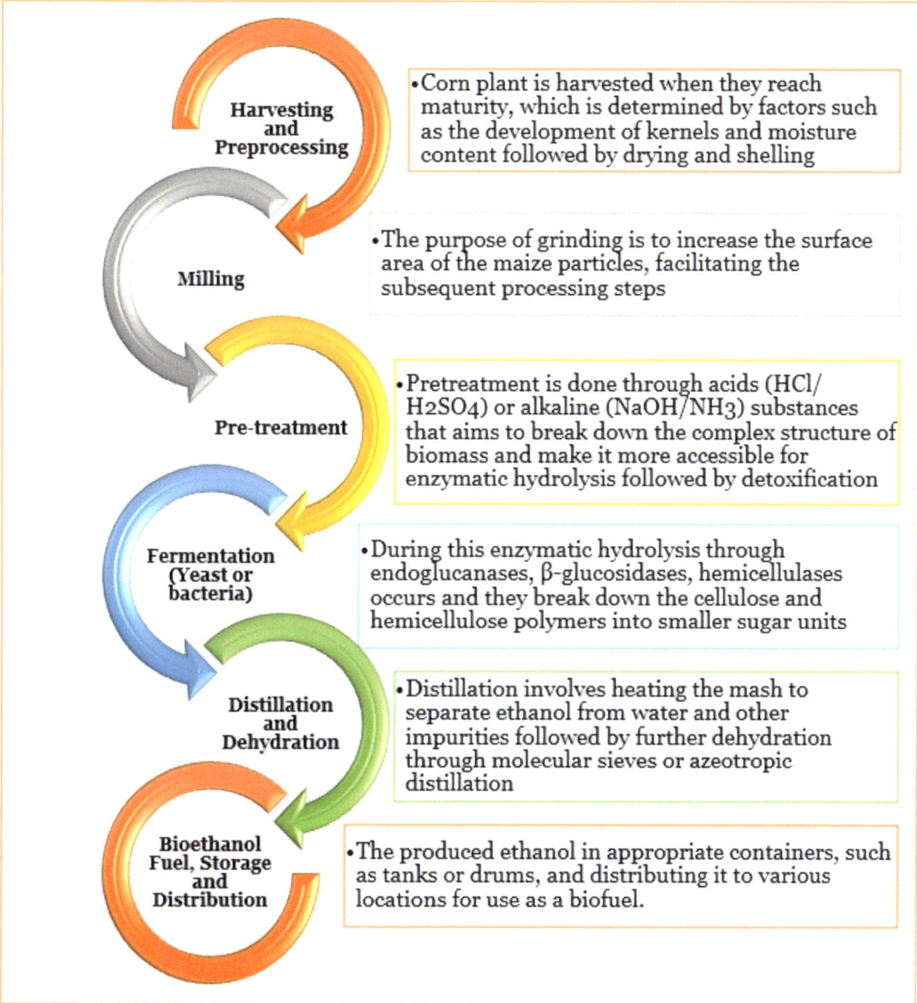

Figure 3.
Steps by step production processes of corn biofuel.

2.3 Production processes and technologies

Corn-based biofuel production processes and technologies are constantly evolving and become progressively popular as a substitute to fossil fuels, due to their renewable and ecologically friendly properties. With this growing demand, there has been a significant increase in research and development efforts aimed at improving corn-based biofuel production processes and technologies. One of the primary technological advances in corn-based biofuel production has been the use of genetically modified (GM) corn that allows for higher yields of ethanol, as the corn has been bred to yield higher amounts of sugars or its relative compounds that can be converted into ethanol. Additionally, new technologies are being developed to enable more efficient conversion of the corn sugars into ethanol, with the aim of reducing production costs and increasing efficiency [21].

Furthermore, there has been an increasing focus on improving the byproducts of corn-based biofuel production. For instance, distillers' grains, which are the leftover byproducts of ethanol production, are being utilized as an animal feed supplement. This not only reduces waste but also provides a source of revenue for biofuel producers [22]. Another futuristic trend is the development of more sustainable and environmentally friendly methods in corn-based biofuel production processes like the use of dry-grind processing, which requires less energy and water than traditional wet-milling methods [23]. Additionally, more sustainable sources of energy are being used to power biofuel production facilities, such as wind and solar power.

3. Prospects of the corn biofuel

Corn-based biofuel offers promising prospects as a renewable energy source. In addition, ongoing plant breeding research, technological advances, and extension approaches and agronomic measures are needed to improve the efficiency and environmental impact of corn-based biofuel production.

3.1 Breeding prospects of corn for biofuel

Breeding corn for biofuel production has been an important research area in recent years. The aim of corn breeding for biofuel is to develop corn varieties that have improved yield and quality traits that can enhance the efficiency of the biofuel production process [24]. It involves the crossing desirable parent plants to produce offspring that have the desired traits with proper evaluation of their performance. The different breeding aspects of corn for biofuel is illustrated below in **Table 1**.

Several breeding techniques can be employed to intensify the generation of biofuel from maize. Here are a few typical methods [26–28]:

3.1.1 Traditional breeding

In traditional breeding, several maize types are crossed to produce hybrids with the appropriate properties for the generation of biofuels. With the use of this technique, breeders can combine advantageous traits like high biomass output, higher sugar or starch content, disease resistance, and stress tolerance.

3.1.2 Marker-assisted selection (MAS)

Marker-assisted selection is a method of breeding that makes use of molecular markers connected to particular qualities of interest. Breeders can selectively breed maize varieties with those markers, expediting the development of desired features, by discovering genetic markers linked to traits related to the generation of biofuels, such as high sugar or starch content [28].

3.1.3 Genomic selection

Genetic information about a person is used to predict that person's performance. Breeders can determine the genetic potential of various individuals for features linked to biofuels by studying the full genome of maize plants. The choice of parent lines for crosses can be influenced by this knowledge, improving the output of biofuel in succeeding generations [29].

Different aspects for breeding	Descriptions
Increased biomass yield	Breeding for corn varieties with higher biomass production, considering traits such as harvest index, stover yield, and overall plant growth.
Enhanced sugar/starch content	Starch is the main component of corn grain, and it is the primary substrate for biofuel production and high starch content in corn can lead to higher ethanol yields, which can make the biofuel production process more cost-effective. Thus, breeding programs aim to develop corn varieties that have higher starch content [5].
Reduced lignin content	Breeding for a lower level of lignin, a compound found in the stalks and leaves of corn plants, can improve the efficiency of the biofuel production process.
Improved conversion efficiency	Breeding for corn varieties with traits that enhance the efficiency of conversion processes, such as increased enzymatic digestibility and higher ethanol yield [25].
Increased oil content	Corn varieties Selection with higher oil content, suitable for biodiesel production, by focusing on oil accumulation and favorable fatty acid profiles [26].
Drought tolerance	Breeding for corn varieties with improved drought tolerance, enabling sustained biomass production under water-limited conditions.
Disease and pest resistance	Incorporating genetic resistance to common corn diseases and pests, reducing yield losses, and ensuring healthier maize crops for biofuel production.

Table 1.
Different aspects of corn breeding for biofuel production.

3.1.4 Genetic engineering

To improve features relevant to biofuel production in maize, certain genes can be inserted using genetic engineering techniques. For instance, genes that boost stress resistance or lower lignin content can be introduced into the genomes of maize plants. Breeders can introduce unique features that might not be present in the maize gene pool thanks to the exact alterations made possible by genetic engineering.

3.1.5 Doubled haploid (DH) method

Homozygous plants with all of their genes present in a single plant are created using the DH method and this strategy can hasten the creation of pure breeding lines with desired features for biofuels. The use of DH technology shortens the breeding process and enables the selection of superior lines based on characteristics including disease resistance, sugar content, and biomass production [26].

3.1.6 High-throughput phenotyping

To quickly and precisely evaluate a variety of plant features, high-throughput phenotyping uses automated methods. Breeders can choose and create better varieties by identifying individuals with superior biofuel-related features using high-throughput phenotyping systems to analyze vast populations of maize plants [27].

3.1.7 Multi-trait selection

Because the generation of biofuels is so complex, it's crucial to take several features into account at once. For which, breeders can apply multi-trait selection techniques that take into account traits like disease resistance, stress tolerance, and nutrient utilization efficiency in addition to biomass output and sugar or starch content. It guarantees a thorough improvement in maize varieties for the production of biofuels.

It's also imperative to note that various breeding techniques are frequently combined and integrated to increase their efficacy. The choice of breeding strategy is also influenced by the resources that are available, the breeding objectives, and the regulatory factors.

3.2 Potential prospects for promoting the adoption of corn as a biofuel crop among farmers

As the biofuel crops provide economic opportunities, environmental sustainability and promotes climate resilience as well as diversified agricultural systems. It is a need of an hour to promote the biofuel crops among farmers. Growing corn as a biofuel crop offers vast opportunities for farmers in income diversification, access to stable markets, value-added production, improved soil health, adoption of conservation practices, government support, and contribution to energy independence [30]. Therefore, it makes corn an important and feasible option for farmers interested in sustainable and economically viable agriculture. The potential prospects for promoting the adoption of corn as a biofuel crop among farmers are as follows.

3.2.1 Educational and skill Upgradation programs

The international conference on research and educational opportunities in biofuel crop production highlighted the importance of education and skill development in bio-fuel crop production to harness the opportunities of entrepreneurship, processing and value addition in the biofuel supply chains among the farmers for income diversification with alternative energy resources [30].

3.2.2 Incentives

Luring the stakeholders with financial incentives such as tax credits, grants, and subsidies to who grow corn for biofuel could be an immediate way but sensitizing the growers to cut down ecological and social cost of production will an ultimate aim for the policy makers to enhance the production as it was achieved in microalgae biofuel production [31].

3.2.3 Advancing market

The marketing sphere of biofuel can be innovatively invented as biofuel shares the potential interaction with food market which is responsible for price hike of food products. Contrary, biofuel production can be serves as potent solution for the food waste and play important role in low carbon economy. These possibilities can be utilized by manufacturers and distributors to develop markets for corn-based biofuels, which can increase demand and provide farmers with a stable market for their crops.

3.2.4 Demonstration trails

The exposure with benefits of growing corn as biofuel as well as provide hands-on training and support for farmers can increase the adoption of this technology among them. As study revealed the robust result in knowledge generation through pilot and demonstration plant on advancing the biofuel innovations in European Union [32].

3.2.5 Extension services

For the production of biofuel crops such as switchgrass, sweet sorghum, miscanthus, soybean, and elephant grass or from micro-algae government provide support program and assistance along with economic incentives [33].

3.2.6 Research and Development

Investment in research for identifying new varieties and efficient management techniques which can be better suited for corn as biofuel production, along with the development of more effective processing methods can foster the adoption of this magical crop as biofuel among farmers. Similar, initiatives had been taken by China for enhancing the adoption of biofuel sugarcane.

3.2.7 Farmer-to-farmer learning

To foster the adoption of biofuel corn the farmer-to-farmer learning networks must be expanded where experienced farmers can share their knowledge and experience with other farmers [34].

3.2.8 Collaborations

For the sustainable transportation of palm biofuel, an international collaboration was done between Malaysia and Colombia. Similar collaborations can with agricultural organizations such as farm bureaus and commodity groups to promote the benefits of growing corn for biofuel.

3.2.9 Social influence

Encourage farmers and stakeholders who have successfully adopted corn as a biofuel crop to serve as advocates and role models for other farmers [35].

3.2.10 Public relations and marketing

Develop public relations and marketing campaigns to raise awareness of the benefits of corn-based biofuels can encourage the farmers to adopt this.

3.3 Best practices for sustainable corn biofuel production

Corn-based biofuels have the potential to contribute significantly to sustainable energy production, provided they are produced using best practices that minimize their negative environmental, social, and economic impacts. Some of the best practices that can be adopted to ensure sustainable corn-based biofuel production (**Figure 4**) are discussed below.

3.3.1 Efficient and sustainable farming practices

It should be firmly adopted to reduce the carbon footprint of corn cultivation which can be achieved through practices such as conservation tillage, crop rotation, and precision agriculture, which help to minimize soil disturbance and reduce the use of fertilizers and pesticides [36–38].

3.3.2 Advanced biofuel production technologies

The usage of advanced biofuel production technologies such as cellulosic ethanol production and waste-to-energy conversion [39] can reduce the pressure on land resources and increase the overall efficiency of the production system.

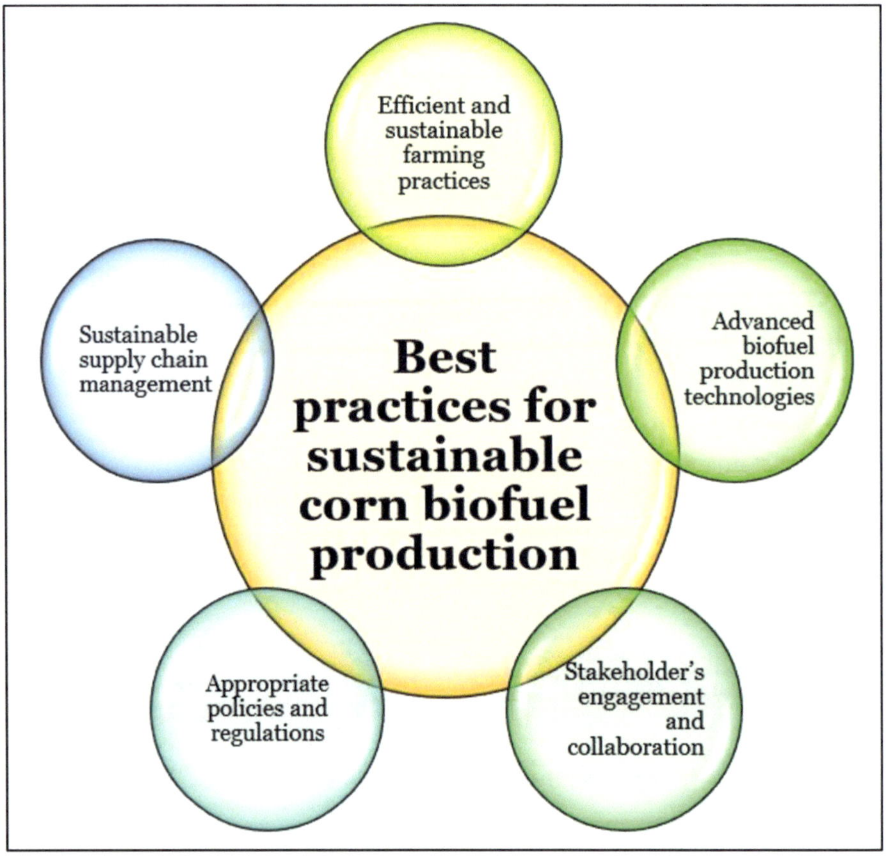

Figure 4.
Best practices for sustainable corn biofuel production.

3.3.3 Sustainable supply chain management

The implementation of sustainable supply chain management practices can ensure that corn-based biofuels are produced and transported in an environmentally and socially responsible manner [40]. This includes the use of sustainable transport modes and the implementation of social and environmental impact assessments throughout the supply chain.

3.3.4 Appropriate policies and regulations

The government policies and regulations that promote sustainable biofuel production should be put in place including incentives for sustainable farming practices, support for the development of advanced biofuel technologies, and regulations that ensure that corn-based biofuels are produced in a sustainable manner [41, 42].

3.3.5 Stakeholder's engagement and collaboration

These are quite essential for promoting sustainable corn-based biofuel production encompassing engaging with all stakeholders including farmers, industry, civil society, and government agencies to promote sustainable biofuel production practices, and to ensure that the concerns of all stakeholders are taken into account [43].

The sustainable production of biofuels requires the adoption of best practices that reduce the negative environmental, social, and economic impacts of biofuel production. By adopting these practices, corn-based biofuels can play a significant role in meeting the growing demand for sustainable energy production.

3.4 Potential benefits and drawbacks of corn-based biofuels

Corn-based biofuels are an increasingly popular alternative to fossil fuels since there are numerous potential benefits associated with the production and use of corn-based biofuels, there are also several drawbacks that must be considered. One of the main benefits of corn-based biofuels is that they can be produced domestically, reducing dependence on foreign oil [44]. Additionally, the production of biofuels creates jobs and provides new revenue streams for farmers [45]. Furthermore, biofuels have the potential to reduce GHG emissions, which can help to mitigate the impacts of climate change [46].

Despite the significant prospects, there are several drawbacks associated with its production and use. One major drawback is that they are often criticized for being less energy efficient than fossil fuels, requiring more energy to produce than they provide in return. This is due to the energy-intensive nature of the production process, which includes planting, harvesting, and processing the corn into biofuel [47]. Also, they compete with food production for land and resources, which can drive up food prices and contribute to food insecurity in developing countries [48]. Furthermore, the expansion of corn biofuel production has been linked to the destruction of natural habitats, such as forests and wetlands, which can have negative impacts on biodiversity [49]. As with any alternative energy source, it is important to carefully weigh the potential benefits and drawbacks before investing in and promoting corn-based biofuel production.

4. Implications of corn-based biofuels

The implications of corn-based biofuel are multifaceted and require careful consideration. While it offers the potential to reduce reliance on fossil fuels and mitigate climate change, there are several notable impacts to consider which are explained as follows.

4.1 Environmental implications

Corn-based biofuels have gained popularity as a renewable alternative to fossil fuels, but their environmental impacts must be carefully considered. The key environmental consideration will be as follow [50].

4.1.1 Loss of biodiversity

The key environmental concern associated with corn-based biofuels is the potential for habitat destruction and biodiversity loss through the conversion of natural habitats to agricultural land for corn production.

4.1.2 Environmental pollution

The use of fertilizers and pesticides in corn production can pollute waterways and harm aquatic ecosystems.

4.1.3 Increased water use

Corn production requires significant amounts of water, and the production of biofuels requires even more water due to the processing and conversion of the corn into fuel. This can exacerbate water scarcity in regions where water resources are already limited [51].

4.1.4 Soil degradation and erosion

The high demand for corn production can lead to intensive farming practices that strip the soil of nutrients and increase the risk of erosion. Soil erosion can contribute to land degradation, loss of agricultural productivity, and increased sedimentation in waterways.

4.1.5 Impact on climate change

While biofuels are often promoted as a way to reduce GHG emissions, the production process itself can be energy-intensive and result in emissions from fertilizer production, transportation, and processing while the life cycle analysis will be taken into consideration [52].

These impacts must be carefully considered when evaluating the viability of corn-based biofuels as a renewable energy source.

4.2 Economic implications

Corn-based biofuels can have both positive and negative economic impacts. Demand for corn as a feedstock for biofuel production can spur growth in the agricultural sector and create jobs in farming, processing, and transportation. In addition, the growth of the biofuels industry can create new employment opportunities in biofuels production and distribution [53]. Additionally, by producing biofuels domestically, countries can reduce their dependence on foreign oil and increase their energy security, while maintaining foreign exchange reserves for other development purposes in the country [41].

However, there are also potential economic drawbacks to consider. Corn biofuel production can contribute to food price volatility and competition with food production for resources. The more corn used for biofuel production, the higher the price of corn as food may be, affecting food prices and food security [48]. In addition, corn-based biofuels can be expensive to produce and use compared to fossil fuels. The production process requires significant energy inputs, and the infrastructure needed to produce and distribute biofuels can be costly [47]. These costs can be passed on to consumers, which can make biofuels less competitive with fossil fuels.

4.3 Social implications

The most imperative social influence is the creation of new jobs at the biofuel processing plant, particularly in rural areas where corn is typically grown which in turn will increase the living standard of the people of those areas. The expansion of the biofuels industry can also spur economic development in these areas and lead to improved infrastructure and services of those communities [54]. In addition, it can also provide environmental benefits by improving air quality, which in turn can have a positive impact on the health of communities near biofuel production facilities [46].

Apart from this, the biofuel production may also lead to land use changes and biodiversity impacts, which can negatively affect local ecosystems and wildlife. Additionally, the benefits of corn-based biofuels may not be evenly distributed across society [49]. For example, the economic benefits may accrue primarily to large agribusinesses, while the negative impacts, such as land use change and biodiversity degradation, may disproportionately affect indigenous communities or marginalized groups. Also, the increasing demand for corn as a feedstock for biofuels may lead to changes in land use practices that impact marginalized communities, potentially leading to displacement of smallholder farmers and land tenure issues [55].

4.4 Policy implications

Corn-based biofuels have gained increasing attention in recent years as a potential solution to reducing GHG emissions and increasing energy security. However, there are several policy implications associated with the production and use of corn-based biofuels. One of the main policy implications is related to the use of government subsidies to support the production of corn-based biofuels. Many governments provide subsidies or tax incentives to encourage the production of biofuels, including corn-based ethanol [56]. However, there is ongoing debate about the effectiveness of these subsidies and their impact on food prices, the environment, and energy security. The use of corn for biofuel production can compete with the production of food crops, leading to potential food price increases and shortages in some areas. Therefore, many governments have implemented policies to limit the amount of corn that can be used for biofuel production, or to encourage the use of non-food crops or agricultural residues as feedstocks. Many governments have implemented sustainability standards and certification schemes to ensure that biofuels are produced in an environmentally and socially responsible manner [42] and in turn aim to promote the use of sustainable practices and avoid negative impacts on land use, biodiversity, and water resources.

The production and trade of biofuels can have significant economic impacts, particularly for countries that are heavily dependent on biofuel exports. Therefore, many governments have implemented trade policies and regulations to manage the international trade of biofuels and ensure fair competition [41].

Another important policy consideration is the likely impact of corn-based biofuels on environmental quality viz., air and water quality. The production and use of biofuels can lead to increased emissions of air pollutants, particularly during the cultivation and processing of feedstocks. Additionally, the production and use of biofuels can have significant water requirements, which can lead to competition with other water users and potential impacts on water quality and availability. Some studies have suggested that the life-cycle emissions of corn-based ethanol may be higher than those of gasoline, particularly when indirect land use changes are considered [57–59]. As a result, there is ongoing debate about the role of corn-based biofuels in meeting climate change goals, and the potential need for alternative biofuel feedstocks and technologies.

Hence, it is imperative to note that the policy implications of corn-based biofuels are complex and interrelated. Policies aimed at promoting the production and use of biofuels may have unintended consequences, particularly if they are not designed and implemented in a coordinated and integrated manner. Therefore, it is essential for policymakers to take a comprehensive and holistic approach to the development of biofuel policies, considering the economic, social, environmental, and technical aspects of biofuel production and use.

5. Conclusions

Corn-based biofuels have played an important role in diversifying the energy mix, reducing GHG emissions, and promoting energy security. Extensive breeding and agronomic research, as well as efforts to expand cultivation, have led to improvements in corn-based biofuel production practices, making them a viable and sustainable option. However, it is important to consider the implications of a heavy reliance on corn for biofuel production. Increased demand for corn as a feedstock can lead to higher prices and potential conflicts with food production. Therefore, it is critical to find a balance between biofuel production and food security to ensure that enough food is available for a growing world population. In addition, the study of corn for biofuels underscores the need for continued innovation and research into alternative feedstocks and advanced technologies. This approach would not only address the potential drawbacks of corn-based biofuels, but also expand the range of sustainable options available. Overall, significant progress has been made in the use of corn as a biofuel, but it is important to take a comprehensive and balanced approach that considers environmental, social, and economic factors. In this way, we can maximize the benefits of corn-based biofuels while minimizing the negative impacts and paving the way for a greener and more sustainable energy future.

Conflict of interest

The authors declare no conflict of interest.

Author details

Smruti Ranjan Padhan[1*], Shankar Lal Jat[2], Pratikshya Mishra[3], Sibananda Darjee[1], Sushmita Saini[1], Soumya Ranjan Padhan[3], Radheshyam[1] and Shivani Ranjan[4]

1 Indian Agricultural Research Institute, New Delhi, India

2 Indian Institute of Maize Research, New Delhi, India

3 Odisha University of Agriculture and Technology, Bhubaneswar, India

4 Dr. Rajendra Prasad Central Agricultural University, Samastipur, India

*Address all correspondence to: smrutiranjaniari@gmail.com

References

[1] Shyam CS, Rathore SS, Shekhawat K, Singh RK, Padhan SR, Singh VK. Precision nutrient management in maize (Zea mays) for higher productivity and profitability. Indian Journal of Agricultural Sciences. 2021;**91**(6):933-935. DOI: 10.56093/ijas.v91i6.114303

[2] FAOStat. Rome: FAO; 2023. Available from: http://www.fao.org/faostat

[3] Grote U, Fasse A, Nguyen TT, Erenstein O. Food security and the dynamics of wheat and maize value chains in Africa and Asia. Frontiers in Sustainable Food Systems. 2021;**4**:617009. DOI: 10.3389/fsufs.2020.617009

[4] Erenstein O, Chamberlin J, Sonder K. Estimating the global number and distribution of maize and wheat farms. Global Food Security. 2021;**30**:100558. DOI: 10.1016/j.gfs.2021.100558

[5] Ragauskas AJ, Williams CK, Davison BH, Britovsek G, Cairney J, Eckert CA, et al. The path forward for biofuels and biomaterials. Science. 2006;**311**(5760):484-489

[6] Morone P, Cottoni L, Giudice F. Biofuels: Technology, economics, and policy issues. In: Handbook of Biofuels Production. Amsterdam, Netherlands: Elsevier; 2023. pp. 55-92. DOI: 10.1016/B978-0-323-91193-1.00012-3

[7] Hoang AT, Pandey A, Huang Z, Nižetić S, Le AT, Nguyen XP. Biofuels an option for agro-waste management. In: Environmental Sustainability of Biofuels. Amsterdam, Netherlands: Elsevier; 2023. pp. 27-47. DOI: 10.1016/B978-0-323-91159-7.00011-4

[8] Sharma A, Pavani Srikavya B, Urade AD, Joshi A, Shyam Narain R, Dwarakanath V, et al. Economic and environmental impacts of biofuels in Indian context. Materials Today. 2023. DOI: 10.1016/j.matpr.2023.05.118

[9] Bhushan S, Jayakrishnan U, Shree B, Bhatt P, Eshkabilov S, Simsek H. Biological pretreatment for algal biomass feedstock for biofuel production. Journal of Environmental Chemical Engineering. 2023;**11**(3):109870

[10] Kumari D, Singh R. Pretreatment of lignocellulosic wastes for biofuel production: A critical review. Renewable and Sustainable Energy Reviews. 2018;**90**:877-891. DOI: 10.1016/j.rser.2018.03.111

[11] Jayakumar M, Gindaba GT, Gebeyehu KB, Periyasamy S, Jabesa A, Baskar G, et al. Bioethanol production from agricultural residues as lignocellulosic biomass feedstock's waste valorization approach: A comprehensive review. Science of the Total Environment. 2023;**879**:163158. DOI: 10.1016/j.scitotenv.2023.163158

[12] Nisa FU. Biofuel: A unique solution for the future energy crisis. In: Environmental Sustainability of Biofuels. Amsterdam, Netherlands: Elsevier; 2023. pp. 219-36. DOI: 10.1016/B978-0-323-91159-7.00008-4

[13] OECD-FAO. OECD-FAO Agricultural Outlook. OECD Agriculture statistics (database). Rome, Italy. 2021. DOI: 10.1787/agr-outl-data-en

[14] Koizumi T. World biofuel continuum: Issues and challenges. In: Environmental Sustainability of Biofuels. Radarweg 29, 1043 NX Amsterdam, Netherlands:

Elsevier; 2023. pp. 69-85. DOI: 10.1016/B978-0-323-91159-7.00009-6

[15] Guo M, Song W, Buhain J. Bioenergy and biofuels: History, status, and perspective. Renewable and Sustainable Energy Reviews. 2015;**42**:712-725. DOI: 10.1016/j.rser.2014.10.013

[16] Morone P, Strzałkowski A, Tani A. Biofuel transitions: An overview of regulations and standards for a more sustainable framework. In: Biofuels for a More Sustainable Futures. Amsterdam, Netherlands. Elsevier; 2020. pp. 21-46. DOI: 10.1016/B978-0-12-815581-3.00002-6

[17] Amer MW, Aljariri Alhesan JS, Ibrahim S, Qussay G, Marshall M, Al-Ayed OS. Potential use of corn leaf waste for biofuel production in Jordan (physio-chemical study). Energy. 2021;**214**:118863. DOI: 10.1016/j.energy.2020.118863

[18] Gu YM, Byun HR, Kim YH, Park DY, Lee JH. Assessing the potential of facile biofuel production from corn stover using attrition mill treatment. Water-Energy Nexus. 2019;**2**:46-49. DOI: 10.1016/j.wen.2020.02.002

[19] Speight JG. Fuels for Fuel Cells. Fuel Cells: Technologies for Fuel Processing. Amsterdam, Netherlands: Elsevier; 2011. pp. 29-48. DOI: 10.1016/B978-0-444-53563-4.10003-3

[20] El-Mashad HM. Biomethane and ethanol production potential of Spirulina platensis algae and enzymatically saccharified switchgrass. Biochemical Engineering Journal. 2015;**93**:119-127. DOI: 10.1016/j.bej.2014.09.009

[21] Maitra S, Singh V. Invited review on 'maize in the 21st century' emerging trends of maize biorefineries in the 21st century: Scientific and technological advancements in biofuel and bio-sustainable market. Journal of Cereal Science. 2021;**101**:103272. DOI: 10.1016/j.jcs.2021.103272

[22] Kurambhatti C, Kumar D, Singh V. Impact of fractionation process on the technical and economic viability of corn dry grind ethanol process. Processes. 2019;**7**:578. DOI: 10.3390/pr7090578

[23] Bothast RJ, Schlicher MA. Biotechnological processes for conversion of corn into ethanol. Applied microbiology and biotechnology. 2005;**67**:19-25

[24] Somerville C. The billion-ton biofuels vision. Science. 2006;**312**:1277

[25] Murray SC, Rooney WL, Mitchell SE, Sharma A, Klein PE, Mullet JE, et al. Genetic improvement of sorghum as a biofuel feedstock: II. QTL for stem and leaf structural carbohydrates. Crop Science. 2008;**48**(6):2180-2193

[26] Himmel ME, Ding SY, Johnson DK, Adney WS, Nimlos MR, Brady JW, et al. Biomass recalcitrance: Engineering plants and enzymes for biofuels production. Science. 2007;**315**(5813):804-807

[27] Sticklen M. Plant genetic engineering to improve biomass characteristics for biofuels. Current Opinion in Biotechnology. 2006;**17**:315-319

[28] Lorenz AJ, Gustafson TJ, Coors JG, De Leon N. Breeding maize for a bioeconomy: A literature survey examining harvest index and Stover yield and their relationship to grain yield. Crop Science. 2010;**50**(1):1-12

[29] Panigrahi KK, Mohanty A, Padhan SR, Guru RKS. Genotoxicity and DNA damage induced by herbicides and toxins in plants. In: Induced Genotoxicity and Oxidative Stress in Plants.

Singapore: Springer; 2021. pp. 29-63. DOI: 10.1007/978-981-16-2074-4_2

[30] Morgan KT, Gilbert RA, Helsel ZA, Buacum L, Leon R, Perret J. White paper report from working groups attending the international conference on research and educational opportunities in bio-fuel crop production. Biomass and Bioenergy. 2010;**34**:1968-1972. DOI: 10.1016/j.biombioe.2010.07.004

[31] Runge CF, Senauer B. Biofuel: Corn isn't the king of this growing domain. Nature. 2007;**450**:478. DOI: 10.1038/450478b

[32] Palage K, Lundmark R, Söderholm P. The impact of pilot and demonstration plants on innovation: The case of advanced biofuel patenting in the European Union. International Journal of Production Economics. 2019;**210**:42-55. DOI: 10.1016/j.ijpe.2019.01.002

[33] Rafa N, Ahmed SF, Badruddin IA, Mofijur M, Kamangar S. Strategies to produce cost-effective third-generation biofuel from microalgae. Frontiers in Energy Research. 2021;**9**:749968. DOI: 10.3389/fenrg.2021.749968

[34] Saini S, Mallick S, Padhan SR. Participatory extension approach: Empowering farmers. Biotica Research Today. 2023;**5**(4):326-328

[35] Gopika R, Fazil TS, Amruth P, Geethalakshmi V, Anandan R, Mathew S. An appraisal of seaweed industry in India at global level: Impact on livelihood and environment and future prospects. Journal of Global Communication. 2021;**14**:7-16. DOI: 10.5958/0976-2442.2021.00002.1

[36] Parihar CM, Jat SL, Singh AK, Ghosh A, Rathore NS, Kumar B, et al. Effects of precision conservation agriculture in a maize-wheat-mungbean rotation on crop yield, water-use and radiation conversion under a semiarid agro-ecosystem. Agricultural Water Management. 2017;**192**:306-319. DOI: 10.1016/j.agwat.2017.07.021

[37] Jat SL, Parihar CM, Singh AK, Nayak HS, Meena BR, Kumar B, et al. Differential response from nitrogen sources with and without residue management under conservation agriculture on crop yields, water-use and economics in maize-based rotations. Field Crops Research. 2019;**236**:96-110. DOI: 10.1016/j.fcr.2019.03.017

[38] Padhan SR, Rathore SS, Prasad SM, Singh RK, Shekhawat K. Influence of nutrient and weed management on weed dynamics and productivity of upland rice (Oryza sativa). The Indian Journal of Agricultural Sciences. 2021;**91**(7):1100-1102. DOI: 10.56093/ijas.v91i7.115141

[39] Lee SY, Sankaran R, Chew KW, Tan CH, Krishnamoorthy R, Chu D-T, et al. Waste to bioenergy: A review on the recent conversion technologies. BMC Energy. 2019;**1**:4. DOI: 10.1186/s42500-019-0004-7

[40] Zarrinpoor N, Khani A. Designing a sustainable biofuel supply chain by considering carbon policies: A case study in Iran. Energy, Sustainability and Society. 2021;**11**:38. DOI: 10.1186/s13705-021-00314-4

[41] Ben-Iwo J, Manovic V, Longhurst P. Biomass resources and biofuels potential for the production of transportation fuels in Nigeria. Renewable and Sustainable Energy Reviews. 2016;**63**:172-192. DOI: 10.1016/j.rser.2016.05.050

[42] Jeswani HK, Chilvers A, Azapagic A. Environmental sustainability of biofuels: A review. Proceedings of the Royal Society A: Mathematical, Physical and Engineering Sciences.

2020;**476**(2243):20200351. DOI: 10.1098/rspa.2020.0351

[43] Leibensperger C, Yang P, Zhao Q, Wei S, Cai X. The synergy between stakeholders for cellulosic biofuel development: Perspectives, opportunities, and barriers. Renewable and Sustainable Energy Reviews. 2021;**137**:110613. DOI: 10.1016/j.rser.2020.110613

[44] Lichtfouse E, Navarrete M, Debaeke P, Souchère V, Alberola C, Ménassieu J. Agronomy for sustainable agriculture: A review. Sustainable agriculture. 2009;**29**:1-7

[45] Chel A, Kaushik G. Renewable energy for sustainable agriculture. Agronomy for Sustainable Development. 2011;**31**:91-118

[46] Mathur S, Waswani H, Singh D, Ranjan R. Alternative fuels for agriculture sustainability: Carbon footprint and economic feasibility. AgriEngineering. 2022;**4**:993-1015. DOI: 10.3390/agriengineering4040063

[47] Malode SJ, Prabhu KK, Mascarenhas RJ, Shetti NP, Aminabhavi TM. Recent advances and viability in biofuel production. Energy Conversion and Management: X. 2021;**10**:100070. DOI: 10.1016/j.ecmx.2020.100070

[48] Kim S, Dale BE. Life cycle assessment of various cropping systems utilized for producing biofuels: Bioethanol and biodiesel. Biomass and Bioenergy. 2005;**29**(6):426-439

[49] Arvanitoyannis IS, Tserkezou P. Corn and rice waste: A comparative and critical presentation of methods and current and potential uses of treated waste. International journal of food science & technology. 2008;**43**(6):958-988

[50] Zhao F, Wu Y, Wang L, Liu S, Wei X, Xiao J, et al. Multi-environmental impacts of biofuel production in the US Corn Belt: A coupled hydro-biogeochemical modelling approach. Journal of Cleaner Production. 2020;**251**:119561

[51] Sheehan JJ. Biofuels and the conundrum of sustainability. Current opinion in biotechnology. 2009;**20**(3):318-324. DOI: 10.1016/j.copbio.2009.05.010

[52] Kauffman N, Hayes D, Brown R. A life cycle assessment of advanced biofuel production from a hectare of corn. Fuel. 2011;**90**(11):3306-3314

[53] Popp J, Harangi-Rákos M, Gabnai Z, Balogh P, Antal G, Bai A. Biofuels and their co-products as livestock feed: Global economic and environmental implications. Molecules. 2016;**21**:285. DOI: 10.3390/molecules21030285

[54] Tenenbaum DJ. Food vs. fuel: Diversion of crops could cause more hunger. Environmental Health Perspectives. 2008;**116**(6):A254-A257. DOI: 10.1289/ehp.116-a254

[55] Barnabe D, Bucchi R, Rispoli A, Chiavetta C, Porta PL, Bianchi CL, et al. Land use change impacts of biofuels: A methodology to evaluate biofuel sustainability. In: Biofuels-Economy, Environment and Sustainability. London, UK: IntechOpen; 2013. DOI: 10.5772/52255

[56] Lapan H, Moschini G. Second-best biofuel policies and the welfare effects of quantity mandates and subsidies. Journal of Environmental Economics and Management. 2012;**63**:224-241. DOI: 10.1016/j.jeem.2011.10.001

[57] Roy P, Tokuyasu K, Orikasa T, Nakamura N, Shiina T. A review of life cycle assessment (LCA) of bioethanol

from lignocellulosic biomass. Japan Agricultural Research Quarterly: JARQ. 2012;**46**(1):41-57

[58] Menten F, Chèze B, Patouillard L, Bouvart F. A review of LCA greenhouse gas emissions results for advanced biofuels: The use of meta-regression analysis. Renewable and Sustainable Energy Reviews. 2013;**26**:108-134

[59] Gallejones P, Pardo G, Aizpurua A, del Prado A. Life cycle assessment of first-generation biofuels using a nitrogen crop model. Science of the Total Environment. 2015;**505**:1191-1201. DOI: 10.1016/j.scitotenv.2014.10.061

Section 5

Modern Ways to Insect Management

Chapter 9

Genetic Delimitation of Fall Armyworm Parasitoids Isolated in Maize in Durango, Mexico

María Berenice González-Maldonado, Miguel Mauricio Correa-Ramírez and Mónica Yazmín Flores-Villegas

Abstract

The fall armyworm *Spodoptera frugiperda* Smith (Lepidoptera: Noctuidae) is the main pest that attacks maize crops in Durango, Mexico. For its biological control, it is desired to use the parasitoids of the Braconidae family; however, its identification is quite complex due to the lack of taxonomic keys that describe the complete morphological characters or are well-defined. It is necessary to study their genetic characters to estimate the variation within populations and species. For this, DNA extraction and amplification by PCR were carried out, as well as the sequencing of a fragment of subunit I of the cytochrome c oxidase (COI) gene. In *Chelonus* sp., morphological variability was observed between *Ch. insularis* and *Ch. sonorensis*, their genetic distances were conspecific, indicating that they probably belong to the same lineage. In *Meteorus*, taxonomically two species were found that had not been reported for Durango: *M. laphygmae* and *M. arizonensis*; however, the genetic distance between these and the species reported in the Genbank® could indicate that it is a single species. These results showed the high morphological and genetic variability in these braconids, probably due to evolutionary and climatic changes.

Keywords: parasitoids, Braconidae, genetic diversity, maize, Mexico

1. Introduction

In Mexico, maize is one of the main agricultural crops. Every year, there is a complex of pests, where Lepidoptera, such as the fall armyworm FAW, stand out. *Spodoptera frugiperda* Smith (Lepidoptera: Noctuidae) is polyphagous to the American continent, where it causes economic losses up to 60% of total yield [1, 2]; among the damages, it causes are loss of photosynthetic area, structural damage to the whorl, direct damage to the grain, and low yields [3]; the pest is controlled with two or three applications of chemical insecticides because it is the most difficult pest to attack, according to an interview conducted by the Dow AgroSciences® company with producers in all regions of the country [4].

In Mexico, despite the importance of growing corn, an integrated pest management program IPM is not used, even though scientists have contributed information in this regard and even less have farmers implemented these techniques together or separately, in mostly only chemical control is used [2, 4].

Failing that, it is currently desired to implement the biological control of this pest due to the environmental benefits that could be achieved with the use of entomopathogenic agents (fungi, bacteria and viruses), some of which are mixed with bioremedial agents [2, 5–7] and natural enemies, where parasitoids stand out, which can be released in a massive way, if their habits, biological cycles and hosts are known, one of the families that has the most potential for the sustainable control of this pest is the parasitoids of the family Braconidae (Hymenoptera: Ichneumonoidea), which represents the parasitoids with the highest taxonomic richness, abundance and distribution in Mexico, after Ichneumonidae [8].

Worldwide, 45 subfamilies, 1103 genera, and 21,221 valid species of Braconidae are recognized [9]. In Mexico, although important attempts have been made to classify and describe species of this family in recent years, only 36 subfamilies, 319 genera, 707 determined species, and 845 morphospecies have been recorded [10].

In Mexico, the Braconidae family is very diverse and abundant in all terrestrial ecosystems; however, of 21,221 species recorded in the world, only 707 species are known. In Mexico, the study of the Braconidae family is extremely important; after Ichneumonidae, it is one of the main families of parasitoids used in the biological control of insects considered pests; they have a great taxonomic richness, and they are regulatory agents of various groups of phytophagous insects: being indicators of the presence or absence of these populations, parasitoids can be used as bioindicator organisms, to monitor changes in an ecosystem affected by anthropogenic activities; in addition, their study helps in understanding the evolution of parasitoid-host interactions, as well as from symbiosis with viruses, and they can be massively released in agriculture and forest environments [11–13].

In Durango, Mexico, for the fall armyworm, the egg, larva, and pupa parasitoids that attack it have been identified; in addition, both taxonomic and genetic studies have been carried out, especially with the Braconidae family, which allows us to know about its diversity and provides a tool to be able to implement biological control measures.

2. The cultivation of maize in Mexico

Maize (*Zea mays*) is native to Mexico, and from the evidence found in Tehuacan, Puebla, it is known that its cultivation began seven thousand years ago. Its domestication allowed the nomadic groups to become sedentary, thus becoming the livelihood of the Mesoamerican peoples.

In Mexico, corn is part of the daily diet, it is the crop with the greatest presence in the country, and it constitutes an input for livestock and for obtaining numerous industrial products, therefore, from the food, economic, political, and social, it is the most important agricultural crop [14].

Mexico is the center of origin of maize. Here, most likely, the greatest diversity of maize in the world is concentrated and here its wild relatives, the teocintles, and another set of related grasses, species of the genus *Tripsacum* (maicillos) have evolved and live [15].

Its production is divided into white and yellow maize; white corn is mainly for human consumption, while yellow corn production is for industry or the manufacture of balanced feed for livestock production.

Corn is the most widely produced maize in the world [4, 16]; the most important countries in terms of planting area in the 2021 agricultural cycle were: China, with an area of 42 million hectares and a production of 273 million tons, followed by the United States and Brazil, with an area of planting of 34.43 and 20.8 million ha and production of 382.6 and 118 million tons, respectively; Mexico ranked sixth in terms of planting area with 7.3 million ha and eighth in terms of production with 28.00 million ha. tons [17]; in Mexico, the main maize-producing states are: Sinaloa, Jalisco, State of Mexico, Guanajuato, and Michoacan [18].

2.1 Main maize pests in Mexico

The primary pests that attack maize are fall armyworm *Spodoptera frugiperda* (FAW), corn earworm *Heliothis zea* (Boddie), blind hen *Phyllophaga* sp., thrips *Frankliniella* sp./*Thrips tabaci* Lindeman [4], maize leafhopper *Dalbulus maidis* (Delong & Wolcott), corn weevil *Geraeus senilis, and* Gyllenhal and *Nicentrites testaceipes* (Champion). The genera and species that appear depend on the region, climatic conditions, and planting season (spring–summer) (winter–spring). Although generally, pests are specific to their host.

2.2 Fall armyworm

The fall armyworm (FAW) (*Spodoptera frugiperda*) has been a consistently important insect pest for several crop species, especially maize, in America for centuries. FAW prefers maize, but it is also common on sorghum and rice and is sporadically important on a vast array of additional crops and plants, including cotton and vegetables [19].

FAW has high fecundity, can rapidly develop resistance to insecticides, and has the capacity to migrate long distances, characteristics which have allowed it to rapidly disperse and establish in different regions (America, Australia, Africa, Asia, E.U. Oceania, Nepal, over 70 countries) [20–22].

2.3 Biological control

Biological control is a component of an integrated pest management strategy. It is defined as the reduction of pest populations by natural enemies, using natural enemies such as parasitoids, predators, pathogens, antagonists, or competitors to suppress pest populations [19].

Biological weed control includes insects and pathogens. Biological control agents for plant diseases are often referred to as antagonists. Parasitoids are species whose immature stage develops on or within a single insect host, ultimately killing the host. Many species of wasps and some flies are parasitoids. Pathogens are disease-causing organisms, including bacteria, fungi, and viruses. They kill or debilitate their host and are relatively specific to certain insect groups [23].

Parasite. It is an organism that lives at the expense of another organism.

Parasitoid. The insect in its immature stage acts as a parasite; when they are adults, they usually fly; parasitoids can kill their host in this case the armyworm. Parasitoids are natural enemies, which are widely used in biological control programs because when an arthropod is parasitized, the female parasitoid inserts its eggs with the help of an ovipositor inside the body of the host or attaches them outside of it, and instead of as long as the pest insect (in this case) continues to develop, it dies and the parasitoid(s) (Diptera and/or Hymenoptera) emerge from its body. The main types of insects that act as parasitoids are wasps, flies, some beetles, mantis flies, and twisted-wing parasites [24].

2.4 Diversity of fall armyworm parasitoids

From ten years to date in Durango, Mexico, studies have been carried out that have made it possible to know the taxonomic diversity of parasitoids of FAW of the families: Ichneumonidae (*Pristomerus spinator* Fabricius, *Campoletis sonorensis* Cameron) [25], Encyrtidae (*Euplectrus plathypenae* Howard), Tachinidae (*Lespesia aletiae* Riley, *L. archippivora* Riley, *Winthemia deilephilae* Osten Sacken, y *Archytas marmoratus* Townsend) [26], Trichogrammatidae (*Trichogramma pretiosum* Riley y *Trichogramma exiguum* Pinto y Platner) and Scelionidae (*Telenomus remus* Nixon) [27], and of the family Braconidae subfamily Homolobinae (*Homolobus truncator* Say) [28], from this same family *Ch. insularis*, *Ch. sonorensis*, *Microchelonus cautus* [29], *M. laphygmae* y *M. arizonensis* [30] the genetic part has also been studied [31, 32].

In the Mexican Republic, the Braconidae family has been studied, even so, there are states where the species are still unknown. **Table 1** shows their distribution in the country.

Parasitoid (genus, species)	State (Mexico)	Autors
Ch. insulares = (Ch. texanus)	Mexico	[33]
	Michoacan	[34]
	Chiapas	[35]
	Chihuahua	[36]
	Veracruz	[34, 37]
	Guanajuato	[38]
	Nayarit	[39]
	Sinaloa	[40]
	Sonora	[41]
	Oaxaca	[42]
	Durango	[43]
	Sinaloa	[44]
Ch. sonorensis	Michoacan	[45]
	Sinaloa	[40]
M. cautus = (Microchelonus cautus)	Sonora	[41]
	Mexico	[33]
	Michaoacan	[45]
	Chiapas	[35]
	Veracruz	[34]
	Nayarit	[39]
	Durango	[43]
Chelonus sp.	Sonora	[41]
	Oaxaca	[42]

Parasitoid (genus, species)	State (Mexico)	Autors
M. arizonensis	Chihuahua	[36]
	Nayarit	[39]
M laphygmae	Michoacan	[45]
	Veracruz	[34, 37]
	Nayarit	[39]
	Sinaloa	[44]
Meteorus sp.	Sinaloa	[40]
	Sonora	[41]

Table 1.
Diversity of parasitoids of the Braconidae family in Mexico.

2.5 Braconidae family

2.5.1 Meteorinae Subfamily

The genus *Meteorus* Haliday (Braconidae: Euphorinae, Meteorini) has 326 globally recorded species from the Nearctic, Neotropical, Palearctic, Oriental, Afrotropical, and Oceanic regions [36, 46]. *Meteorus* is a cosmopolitan genus of koinobiont endoparasitoids of Coleoptera and Lepidoptera [47, 48]; *Meteorus* is a paraphyletic group and its rearrangement into several monophyletic genera is pending [47]. The mature larvae of some species spin a cocoon suspended by a thread, and it is from this habit that the name of the genus is derived [36].

2.5.2 Cheloninae Subfamily

Cheloninae Förster is a moderately large subfamily within the family Braconidae. The subfamily comprises more than 1500 described species in the world. Members of this subfamily are present in almost all geographic regions [29, 49].

2.6 Mexico distribution

2.6.1 Chelonus

In 1995, for Guanajuato (state in the center of the country) the genera *Ascogaster*, *Chelonus,* and *Phanerotoma* of the subfamily Cheloninae were reported. Over the years, studies on the taxonomy of this genus have increased and more is known about its diversity.

For Mexico, it has been reported to *Chelonus busckiella* Viereck, 1912; *Ch. davinervis* Cameron, 1904; *Ch. insulares* Cresson, 1865; *Ch. mexicanus* Brètes, 1927; *Ch. quadrimaculatus* Cameron, 1887; *Ch. sericeus* Say, 1824, *Ch. sonorensis* Cameron, 1887; *Microhelonus blackburni* Cameron, 1886; *Microchelonus cautus* (Cresson, 1872), *M. heliopae* Gupta, 1955; *M. pectinophorae* Cushman, 1931 y *M. phrhorimaeae* Gahan, 1917 [9].

In the state of Durango, Mexico, it has been reported to *Chelonus insulares*, *Ch. sonorensis* y *Ch. cautus* (= *Microchelonus*); however, based on the coloration patterns in the metasoma (irregular spots), eight morphotypes were found that did not match the taxonomic keys of [50–52]; therefore, its molecular identification was necessary; in this regard, [53] identified seven species of fall armyworm parasitoids, including *Ch. insularis* (isolates from Colima, Jalisco), *Ch. cautus* (Colima, Puebla, Nayarit) and *M. laphygmae* (Puebla, Colima)., using polymerase chain reaction amplification and restriction enzyme digestion, this enables the precise determination of the species of those parasitoids larvae that are usually not morphologically identifiable, where they appeared equal size amplification of the cytochrome c oxidase subunit 1 fragment was obtained for all seven species. It is also recommended to carry out genitalia or morphometry studies.

2.6.2 Meteorus

In 1990, in Tamaulipas and Nuevo Leon, Mexico, it was reported to *Meteorus* prob. *laphygmae*, *M.* prob. *versicolor,* and four more species of *Meteorus*, unknown up to that time [48]. Other species of *Meteorus* have been reported over time, but there are few studies regarding their genomic sequences.

In Durango, Mexico, the genus *Meteorus* is mostly distributed in Santiago, Papasquiaro, and Durango, probably due to variations in climate and altitude (**Figure 1**). It belongs to the region of Las Quebradas and the other municipalities to the region of valleys and smooths.

From 2012 to date, studies have been carried out in various locations in municipalities located in the center and north of the state of Durango, which has allowed us to know the diversity of fall armyworm parasitoids. **Figure 1** shows the sampled municipalities.

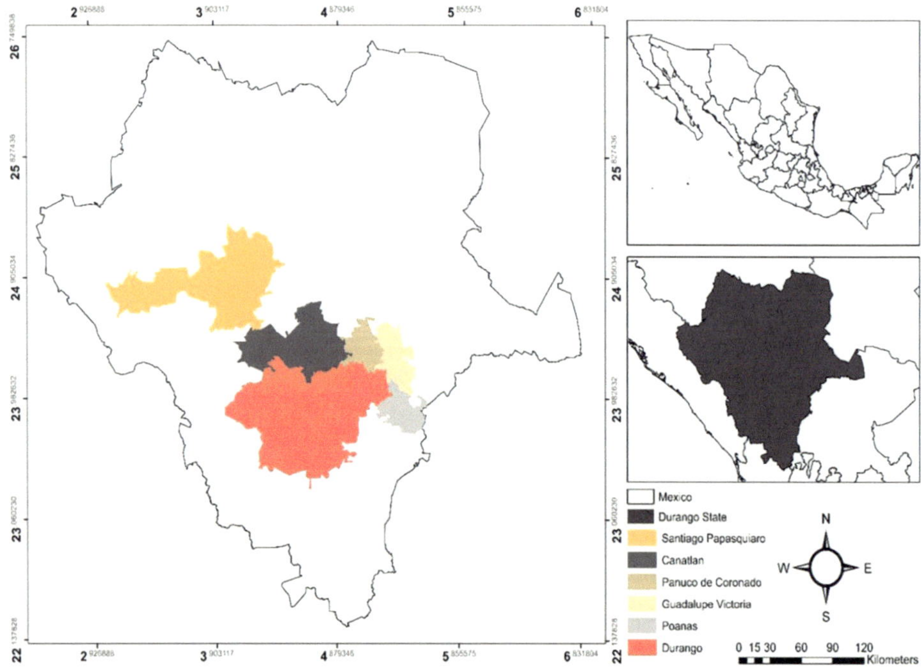

Figure 1.
Distribution of the Braconidae subfamily in Durango, Mexico.

Ch. insularis is the parasitoid that is mostly distributed in Durango and Mexico [54]; however, in the last two years, in Durango *Meteorus* sp., it is the parasitoid that presents greater capacities to be massively reproduced in the laboratory due to its development on an artificial diet (data not yet published).

2.7 Morphological delimitation

2.7.1 *Chelonus*

Specimens with morphological characters to belong to this genus were separated using the taxonomic keys of [50]. Species identification was carried out by PhD. Alejandro Gonzalez-Hernandez, through the comparison of the preserved material with reference specimens from the Collection of Entomophagous Beneficial Insects of the Facultad de Ciencias Biologicas de la Universidad Autonoma de Nuevo Leon, Mexico.

2.7.2 *Meteorus*

The obtained parasitoids were labeled and preserved in 70% alcohol. The Meteorinae (Euphorinae) material was studied at the Insect Museum (MI-FA) of the Universidad Autonoma de Tamaulipas, where it was mounted and labeled using the EntoPrint program with the respective collection data. For the determination of the subfamily and genus, we used the keys of [50] while for the determination of the species the keys of [55, 56]. For Durango, Dgo., Mexico, it has been reported to *Meteorus arizonensis* Muesebeck (Hymenoptera: Braconidae) y *Meteorus laphygmae* Viereck (Hymenoptera: Braconidae); however, their morphological characters do not coincide 100% with the taxonomic keys because they present color patterns in the mesosome that could indicate that they are other species. In this regard, [57] pointed out upon the unreliable color variability in identifying species. In fact, the color pattern is a variable that might be affected by environmental conditions [58].

The color patterns in *M. arizonensis* and *M. laphygmae* should not be considered as distinctive to identify a species; in *Meteorus,* there were nine morphotypes or different color patterns in the mesosome of the specimens of this species, even so, genetically they all belong to the same species; however, in this regard, [59] indicate that this property (melanism) increases the flight activity of wasps at low ambient temperatures of *M. pulchicornis* (Wesmael); they subjected this parasitoid (coconuts) to different temperatures (15, 20, 25 and 30°C); it was observed that at the lowest temperatures, the body of the parasitoid darkened more, which could indicate that the color change in some morphological characters of the parasitoids is due to the change in their body temperature and the environment in which they develop and not that they are different species. Similar situation with *Chelonus*, where coloration patterns in the metasome indicate that *Ch. insularis* and *Ch. sonorensis* belong to the same species [29, 32].

2.8 Genetic delimitation

2.8.1 DNA extraction

Twenty-seven individuals belonging to *Meteorus laphygmae* with five individuals, followed by *M. arizonensis* with four specimens, *Chelonus insularis* with 14 specimens,

Ch. sonorensis with two specimens, and *Ch. cautus* with three specimens, separated according to their morphological characteristics, were used.

Total genomic DNA was isolated per individual using the Promega DNA extraction kit, following the manufacturer's instructions with some modifications. Briefly, the digestion time was modified taking a total of 16 h at 56°C in a dry bath with continuous shaking. The next step was cleaning the aqueous phase with salt precipitation of detergent, proteins, and lipids followed by an organic solvent cleanup (adding 350 μl of chloroform-isoamyl alcohol 24:1), mixed it by inversion for 20 s each sample and centrifuge it by five minutes to recover the aqueous phase in a new tube (1.5 ml).

The DNA precipitation was reached by adding 1.5 volumes of cool isopropanol, followed by storing samples at −20°C for 12–16 h. Samples were cleaned with cool ethanol 80% two times. After washing samples with ethanol, those were dried and hydrated with 60 μl of milli-Q water.

From the isolated DNA, a fragment of the mitochondrial cytochrome c oxidase I (COI) gene was amplified in the individuals of the species analyzed using oligonucleotides HCO-2198 (5′-TAA ACT TCA GGG TGA CCA AAA AAT CA-3′) (forward) and LCO-1490 (5′-GGT CAA CAA ATC ATA AAG ATA TTG G-3′) (reverse) reported by [60]. PCR conditions were 1 min 30 s at 94°C, denaturation 35 cycles at 94°C (1 min), alignment at 50°C (1 min), extension at 72°C for 1 min, and a final extension step at 72°C (15 min) in a thermal cycler (Model 9600, Labnet International, Edison, NJ) using 50–150 ng of DNA, 0.40 pmol of each oligonucleotides, 2.5 mM of MgCl 2, 0.2 mM of each dNTPs (Promega, Madison, WI), 1× of polymerase chain reaction buffer, and 1 unit of Taq polymerase (Promega) in a final volume of 50 μl [53, 61].

Amplified PCR products in 1% agarose gels stained with ethidium bromide were visualized by electrophoresis and observed at 430 nm in a UV transilluminator. The double-chain products were purified using a Wizard SV Gel and PCR Clean-up purification system. The amplified products were sequenced on a Genetic Analyzer Applied Biosystems 310 using the method of big dye terminator (Applied Biosystems Inc., Foster City, CA). The sequence files were edited and aligned using Chromas Pro ver. 2.1.10.1.

The sequences were translated into proteins to confirm the identity of the fragments [62]. Multiple alignments used Clustal X [63] with gap opening costs = 50, gap extension = 6.6, divergent delay of sequences = 30%, and DNA transition weight = 0.5 [64]. Genetic diversity in the species was measured as haplotype diversity (h), number of private haplotypes (P), and nucleotide diversity (p) analyzed using Arlequin version 3.5.1.21 and DnaSP version 5.1 [65]. Arlequin version 3.5.1.21 [66] was used for analysis of molecular variance (AMOVA) of population structure. Genetic differences between individuals were analyzed. Sum of the squares of deviation (SSD) and index of Harpending's-Raggedness [67] were calculated to evaluate the fit of the observed data using a model of sudden demographic expansion or a model of geographic range expansion. The mismatch distribution was compared with expected distributions by models of sudden population expansion [68] and spatial expansion [69, 70].

To reconstruct phylogenetic relationships, we used Bayesian inferences. We used *Campoletis sonorensis* Cameron, *C. flavicincta* (Ashmead), and *Homolobus truncator* Say as outgroups to polarize the characters within samples of *Meteorus* spp. and *Chelonus* spp.

Bayesian analysis used the GTR model with invariant rate heterogeneity. A posterior probability analysis [71] was performed using the program MrBayes version 3.0b4 [72]. Bayesian posterior probability calculations were implemented in a range of ten million generations, sampling every 1000 generations, and discarding the first

1000 trees sampled (as burn-in). Support for nodes was determined by posterior probabilities [72, 73]. For Bayesian analysis, three independent runs were conducted to get an impression of the robustness of the phylogenetic reconstruction.

The amplified fragments were between 695 and 710 bp, the sequences were editing and adjusting them to 650 bp. Regarding the AMOVA, the maximum distribution of variance was observed when two groups were formed: group 1: *Ch. insularis* + *Ch. sonorensis* and group 2: *Ch. cautus*; this indicates that the groups are different from each other (FSC = 0.02289, p = 0.011 at 95% confidence), there is no difference within them (FST = 0.97679, p = 0.43 at 95% confidence) (**Figures 2** and **3**), these were compared with group 3: *Ch. insularis* from different states of Mexico, the key assigned in the GeneBank® is indicated [53].

The unpaired distribution of DNA (Mismatch distribution) showed two peaks that reinforce the existence of two groups corresponding to group 1 and group 2, and each of them presented a sudden population increase.

2.8.2 Phylogeny

The median-joining networks of *Meteorus* and *Chelonus* haplotypes did not reveal divergent clusters of haplotypes by phenotype or color. Rather, the COI networks are star-shaped [74], whereas the network shows no structure, indicating that the phylogenetic information given by these sequences is adequate for phylogenetic inference. The trees generated by the Bayesian analyses are mostly unresolved within analyzed species, and the clusters that are formed may contain sequences from different localities and different morphospecies.

The phylogenetic affinities showed that the analyzed specimens of *Ch. insulares* and *Ch. sonorensis* are in a single group, where both are mixed (without forming different groups); in turn, this group is separated from another group of sequences belonging to *Ch. insularis* from other parts of Mexico. In the case of *Ch. cautus*, it is observed that the sequences obtained in this work tend to form a single group that is separated from

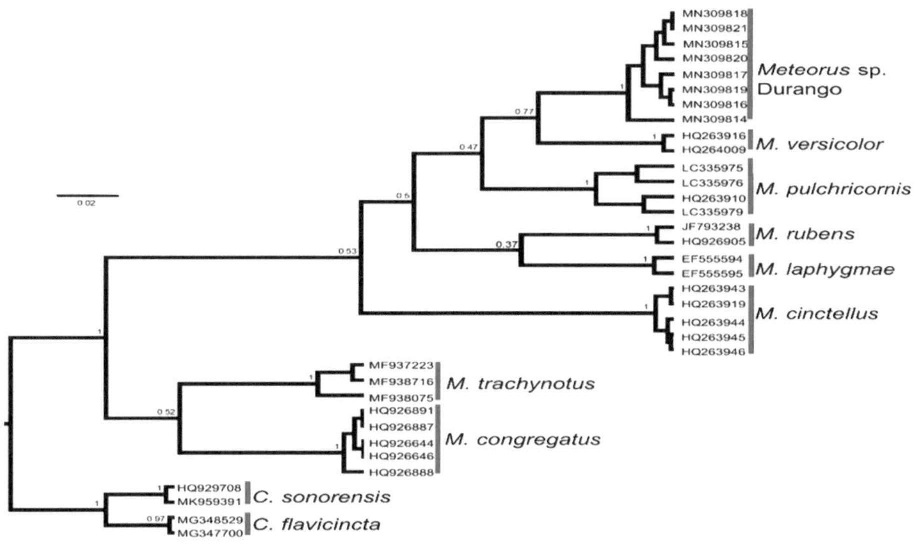

Figure 2.
Phylogenetic tree showing the groups of Meteorus sp. de Durango, compared to other species.

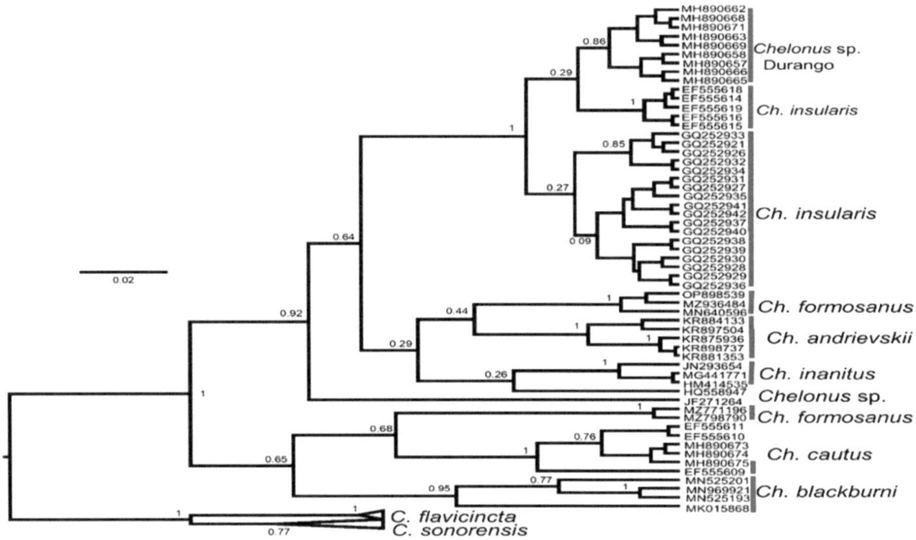

Figure 3.
Phylogenetic tree showing the groups of Chelonus sp. de Durango, compared to other species.

Chelonus insularis + *Ch. sonorensis;* it is reinforced by genetic distances between the two morphospecies of *Chelonus* (G. D. = 0.005) that occur in Durango, Mexico suggests that there is a reproductive isolation among populations that occurs in central and southern parts of Mexico contrasted with species from Durango. In the case of *M. laphygmae* and *M. arizonensis,* the phylogenetic affinities are like that described earlier, it means that the two morphospecies of *Meteorus* from Durango form a single genetic group or *M. laphygmae* + *M. arizonensis* represents a single species with a genetic distance closed to zero (G. D. = 000001) and the genetic distances with other valid species of the same genus are up to 5%, as shown in **Table 2** (G. D. higher than 0.050).

3. Discussion

Mitochondrial DNA fragment (COI) presented a high amount of genetic diversity. The high genetic diversity found is unlikely to be due to sequencing error or artifact because sequences were run in both directions, and we sequenced five samples to verify consistency of the results. We did not detect heterozygote base calling (double peaks in any sequence direction) in COI fragment. Furthermore, the COI data set was checked by amino acid translation, and we found no stop codons within sequences. Notwithstanding these high levels of variation, we were unable to detect any structure in the data. We expected that the pattern of genetic variation would reflect that of morphological variation shown us. We found no relationship between morphological variation and genetic variation.

4. Conclusions

This study has allowed to know the species of parasitoids of the family. Braconidae of fall armyworm, an important pest of maize in Durango, which

Species A	Species B	Gen. Dist.
Meteorus sp *Durango*	*M. versicolor*	0.058
Meteorus sp Durango	*M. rubens*	0.103
Meteorus sp Durango	*M. laphygmae*	0.118
Meteorus sp Durango	*M. pulchricornis*	0.112
Meteorus sp Durango	*M. cinctellus*	0.146
Meteorus sp. Durango	*M. sp.congregatus*	0.184
Meteorus sp. Durango	*M. trachynotus*	0.264
Meteorus sp. Durango	*Campoletis sonorensis*	0.284
Meteorus sp. Durango	*Campoletis flavicincta*	0.276
Chelonus sp. Durango	*Ch. blackburni*	0.109
Chelonus sp. Durango	*Ch. inanitus*	0.128
Chelonus sp. Durango	*Ch.andrievskii*	0.133
Chelonus sp. Durango	*Ch. formosanus*	0.134
Chelonus sp. Durango	*Chelonus* sp.	0.135
Chelonus sp. Durango	*Ch. cautus* Dgo.	0.160
Chelonus sp. Durango	*Ch. cautus*	0.164
Chelonus sp. Durango	*Campoletis sonorensis*	0.319
Chelonus sp. Durango	*Campoletis flavicincta*	0.318

Table 2.
Genetic distances between valid species of braconid wasps versus species that occurs in Durango, Mexico. Gen. Dist., Genetic distances calculated by DNA mutation model of Kimura 2 Parameters.

presented genetic variability, taxonomically their characteristics coincide with the diagnoses established for hundreds of years; however, genetically not. The study does not contemplate its redescription, but it provides important aspects of its genetic characterization, mainly of the genera *Chelonus* and *Meteorus*; these parasitoids contribute to the biodiversity of hymenopteran parasitoids of this pest in the corn region of Durango, which can also be candidates within the biological control of the pest within a context of sustainability and good agricultural practices, contributing to the environment. The delimitation of parasitoid species using taxonomic tools combined with the use of a molecular characterization allowed to clarify taxonomic hypotheses and doubts.

Author details

María Berenice González-Maldonado[1*], Miguel Mauricio Correa-Ramírez[1] and Mónica Yazmín Flores-Villegas[2]

1 Instituto Politécnico Nacional, CIIDIR IPN Unidad Durango, Mexico

2 Universidad Politécnica de Durango, Mexico

*Address all correspondence to: monica.flores@unipolidgo.edu.mx

References

[1] INECOL. Instituto de Ecologia, A.C. El gusano cogollero del maiz una plaga que viaja por el mundo [Internet]. 2023. Available from: https://www.inecol.mx/inecol/index.php/es/ct-menu-item-25/ct-menu-item-27/17-ciencia-hoy/1194-el-gusano-cogollero-del-maiz-una-plaga-que-viaja-por-el-mundo [Accessed: June 02, 2023]

[2] Vazquez-Ramirez MF, Rangel-Nuñez JC, Ibarra JE, Del Rincon-Castro MC. Evaluacion como agentes de control biologico y caracterizacion de cepas mexicanas de *Bacillus thuringiensis* contra el gusano cogollero del maiz *Spodoptera frugiperda* (Lepidotera: Noctuidae). Interciencia. 2015;**40**:397-402. DOI: 0378-1844/14/07/468-08

[3] Lopez J, Chirinos D, Ponce W, Solorzano R, Alarcon J. Actividad insecticida de formulados botanicos sobre el gusano cogollero, *Spodoptera frugiperda* (Lepidoptera: Noctuidae). Revista Colombiana de Entomologia. 2022;**48**:1-8. DOI: 10.25100/socolen.v48i1.11739

[4] Blanco CA, Pellegaud JG, Nava-Camberos U, Lugo-Barrera D, Vega-Aquino P, Coello J, et al. Maize pests in Mexico and challenges for the adoption of integrated pest management programs. Journal of Integrated Pest Management. 2014;**5**:E1-E9. DOI: 10.1603/ipm14006

[5] Pacheco-Hernandez MA, Resendiz-MartInez JF, Arriola-Padilla VJ. Entomopathogenic organisms for pest control in the Mexican agriculture, livestock, and forest sectors: A review. Revista Mexicana de Ciencias Forestales. 2019;**10**:1-29. DOI: 10.29298/rmcf.v10i56.496

[6] Del Rincon Castro MC, Mendez Lozano J, Ibarra JE. Caracterizacion de cepas nativas de *Bacillus thuringiensis* con actividad insecticida hacia el gusano cogollero del maiz *Spodoptera frugiperda* (Lepidoptera: Noctuidae). Folia Entomologica Mexicana. 2006;**45**:157-164. Available from: Redalyc.org/42445206

[7] Martinez-Castillo AM, Pineda-Ortega S, Samuel-Pineda G, Figueroa de la Rosa JM, Palma-Castillo LJ, Ramos-Ortiz S. Combinacion de un baculovirus con compuestos biorracionales para el control de gusano cogollero. Revista Entomologia Mexicana. 2021;**8**:1-7

[8] Coronado-Blanco JM, Zaldivar-Riveron A. Biodiversity of Braconidae (Hymenoptera: Ichneumonoidea) in Mexico. Revista Mexicana de Biodiversidad, Supl. 2014;**85**:S372-S378. DOI: 10.7550/rmb.32000

[9] Yu DS, Van Achterberg K, Horstmann K. Taxapad 2012 - World Ichneumonoidea 2011. In: Taxonomy, Biology, Morphology, and Distribution. Canada. On CD. Vancouver; 2012. Available from: http://www.taxapad.com

[10] Coronado-Blanco JM. La familia Braconidae (Hymenoptera) en Mexico. Entomologia Mexicana. 2013;**12**:31-44

[11] Rodríguez JL, Flores-Davila M, Coronado-Blanco JM, Garcia-Martinez O, Aguirre-Uribe LA. Contribucion a la familia Braconidae (Hymenoptera: Ichneumonidae) del Estado de Aguascalientes, Mexico. Acta Zoologica Mexicana. 2019;**35**:1-10

[12] Figueroa JI, Mejia-Ramirez A, Martinez AM, Pineda S, Ponce-Saavedra J, Sanchez-Garcia JA, et al. Diversidad de especies de *Blacus* (Hymenoptera: Braconidae, Blacinae) en dos ecosistemas forestales de Michoacan, Mexico. Acta Zoológica Mexicana (nueva serie). 2021;**37**:1-15. DOI: ht10.21829/azm.2021.3712316

[13] Sulca Garro LA. Distribución de Rogadinae (Hymenoptera: Braconidae) en una gradiente altitudinal en los Andes del Sur de Peru. Ecología Aplicada. 2013;**12**:141-146

[14] Heinrich Boll Stiftung. México: Un País de Maíz [Internet]. 2023. Available from: https://mx.boell.org/es/2019/04/17/mexico-un-pais-de maiz#:~:text=Es%20la%20base%20de%20alimentaci%C3%B3n,principalmente%20su%20almid%C3%B3n%20y%20aceite [Accessed: June 25, 2023]

[15] Biodiversidad Mexicana. Maices [Internet]. 2023. Available from: https://www.biodiversidad.gob.mx/diversidad/alimentos/maices [Accessed: May 03, 2023]

[16] USDA. United State Department of Agriculture. Feed Grains Sector at a Glance [Internet]. 2023. Available from: https://www.ers.usda.gov/topics/crops/corn-and-other-feed-grains/feed-grains-sector-at-a-glance/#:~:TEXT=The%20United%20States%20is%20the, grown%20in%20the%20Heartland%20region. [Accessed: June 15, 2023]

[17] USDA. United State Department of Agriculture. World Agricultural Production. United State Department of Agriculture, Foreign Agricultural Service (USDA-FAS). Circular Series WAP 11-16. Available from: https://apps.fas.usda.gov/psdonline/circulars/production.pdf [Accessed: June 02, 2023]

[18] Gobierno de Mexico. Maiz, cultivo de Mexico [Internet]. 2023. Available from: https://www.gob.mx/agricultura/articulos/maiz-cultivo-de-mexico#:~:text=Cultivo%20de%20ma%C3%ADz%20en%20M%C3%A9xico,los%2032%20estados%20del%20pa%C3%ADs [Internet]. 2023. [Accessed: June 02, 2023]

[19] Hruska AL. Fall armyworm (*Spodoptera frugiperda*) management by smallholders. CAB Reviews. 2019;**14**:1-11. Available from: http://www.cabi.org/cabreviews

[20] Agarwal A, Rako L, Schutze MK, Starkie ML, Tek-Tay W, Rodoni BC, et al. A diagnostic LAMP assay for rapid identification of an invasive plant pest, fall armyworm *Spodoptera frugiperda* (Lepidoptera: Noctuidae). Scientific Reports. 2022:1116

[21] Shree Ratna BA, Binu B, Premnidhi S, Pathour S, Naresh M, Tahseen H. First record of fall army worm *Spodoptera frugiperda* (J. E Smith) from Nepal. Indian Journal of Entomology. 2019;**81**:635-639. DOI: 10.5958/0974-8172.2019.00137.8

[22] USDA. United State Department of Agriculture. Feed Grains Sector at a Glance [Internet]. 2023. Available from: https://www.ers.usda.gov/topics/crops/corn-and-other-feed-grains/feed-grains-sector-at-a-glance/#:~:text=The%20United%20States%20is%20the,grown%20in%20the%20Heartland%20region [Accessed: June 14, 2023]

[23] USDA. United State Department of Agriculture. Biological Control Program [Internet]. 2023. Available from: https://www.aphis.usda.gov/aphis/ourfocus/planthealth/plant-pest-and-disease-programs/biological-control-program/

ct_biological_control_program [Accessed: May 05, 2023]

[24] Utah. General Concepts of Biological Control [Internet]. 2023. Available from: https://extension.usu.edu/pests/research/biological-control-concepts Accessed: May 18, 2023]

[25] Garcia-Gutierrez C, Gonzalez-Maldonado MB, Gonzalez-Hernandez A. Parasitismo natural de Braconidae e Ichneumonidae (Hymenoptera) sobre *Spodoptera frugiperda* (Lepidoptera: Noctuidae). Revista Colombiana de Entomologia. 2013;**39**:25-29

[26] Gonzalez-Maldonado MB, Hernandez-Zetina DA, Ruiz-Cancino E. Parasitoides (Diptera: Tachinidae) del gusano cogollero *Spodoptera frugiperda* J. E. Smith (Lepidoptera: Noctuidae) en maiz en Durango, Mexico. Southwestern Entomologist. 2018;**43**:183-187. DOI: 10.3958/059.043.0110

[27] Gonzalez-Maldonado MB, Garcia-Gonzalez F, Ramirez-Ahuja ML, Chairez-Hernandez I, Flores-Villegas MY. Microhymenoptera parasitoids of corn earworm in Maize at Durango, Mexico. Southwestern Entomologist. 2022;**47**:887-893. DOI: 10.3958/059.047.0412

[28] Gonzalez-Maldonado MB, Coronado-Blanco JM. A New record of *Homolobus* (*Apatia*) *truncator* (Say) in Durango, Mexico. Southwestern Entomologist. 2020a;**45**:319-323. DOI: 10.3958/059.045.0137

[29] Gonzalez-Maldonado MB, Coronado-Blanco JM, Chairez-Hernandez I, Correa-Ramirez MM. Delimitacion morfologica entre *Chelonus insularis* y *Chelonus sonorensis* en la region maicera de Durango, Mexico. Southwestern Entomologist. 2021;**46**:765-772. DOI: 10.3958/059.046.0318

[30] Gonzalez-Maldonado MB, Coronado-Blanco JM, Lomeli-Flores JR. Nuevos registros de bracónidos (Hymenoptera: Braconidae) parasitoides de gusano cogollero, *Spodoptera frugiperda* (Lepidoptera: Noctuidae) en Durango, México. Revista Colombiana de Entomología. 2020b;**6**:1-4. DOI: 10.25100/socolen.v46i2.8435

[31] Gonzalez-Maldonado MB, Correa-Ramirez M, Rosas-Garcia NM, Chairez-Hernandez I, Garzon-Zuñiga MA. Genetic variability of species of the genus *Meteorus* Haliday, 1935, at Durango, Mexico. Southwestern Entomologist. 2019a;**44**:909-918. DOI: 10.3958/059.044.0406

[32] Gonzalez-Maldonado MB, Correa-Ramirez MM, Mireles-MartinezM,Chairez-HernandezI, Garzon-Zuñiga MA. Analisis de secuencias del gen del citocromo oxidasa c region I sugiere una alta cercania genetica entre *Chelonus insularis* y *Chelonus sonorensis*. Southwestern Entomologist. 2019;**44**:303-306. DOI: 10.3958/059.044.0132

[33] Molina OJ, Carpenter JE, Heinrichs EA, Foster JE. Parasitoids and parasites of *Spodoptera frugiperda* (Lepidoptera: Noctuidae) in the Americas and Caribbean basin: An inventory. Florida Entomologist. 2003;**86**:254-289

[34] CIMMYT. Se identifican numerosos insectos que combaten al gusano cogollero [Internet]. Available from: https://www.cimmyt.org/es/noticias/se-identifican-numerosos-insectos-que-combaten-al-gusano-cogollero/ [Accessed: June 09, 2023]

[35] Ruiz-Najera RE, Molina-Ochoa J, Carpenter JE, Espinosa-Moreno JA,

Lezama-Gutierrez R, Foster JE. Survey for hymenopteran and dipteran parasitoids of the fall armyworm (Lepidoptera: Noctuidae) in Chiapas, Mexico. Journal of Agricultural and Urban Entomology. 2007;**24**:35-42. DOI: 10.3954/1523-5475-24.1.35

[36] Ordoñez GM, Bustillos RJ, Loya MJ, Rios VC, Jacobo CJ. Parasitoides de *Spodoptera frugiperda* (J. E. Smith) (Lepidoptera: Noctuidae) en Chihuahua, Mexico. Metodos en Ecologia y Sistematica. 2015;**10**:66-72

[37] Hoballah M, Degen T, Bergvinson D, Savidan A, Tamo C, Turlings T. Occurrence and direct control potential of parasitoids and predators of the fall armyworm (Lepidoptera: Noctuidae) on maize in the subtropical lowlands of Mexico. Agricultural and Forest Entomology. 2004;**6**:83-88

[38] Arroyo-Maciel GA, Salas AM, A. Enemigos naturales de *Spodoptera frugiperda* y *Helicoverpa zea* como una alternativa al uso de insecticidas. Jovenes en la ciencia. Revista de divulgación científica. 2015;**1**:48-53

[39] Gutierrez-Ramirez A, Robles-Bermudez A, Cambero-Campos J, Coronado-Blanco JM. *Meteorus arizonensis* Muesebeck, 1923 (Hymenoptera: Braconidae): New record for Mexico. Acta Zoologica Mexicana. (n.s.). 2015;**31**:123-124

[40] Cortez-Mondaca E, Perez-Marquez J, Bahena-Juarez F. Control biologico natural de gusano cogollero (Lepidoptera: Noctuidae) en maiz y en sorgo, en el norte de Sinaloa, Mexico. Southwestern Entomologist. 2012;**37**:423-428

[41] Cortez-Mondaca E, Armenta-Cardenas I, Bahena-Juarez F. Parasitoides y porcentaje de parasitismo sobre el gusano cogollero *Spodoptera frugiperda* Smith (Lepidoptera: Noctuidae) en el Sur de Sonora, Mexico. Southwestern Entomologist. 2010;**35**:199-203

[42] Cruz-Sosa E. Evaluacion del parasitismo natural de Spodoptera frugiperda (Smith) (Lepidoptera: Nocttuidae) [thesis]. CIIDIR-IPN Unidad Oaxaca; 2009. Available from: http://literatura.ciidiroaxaca.ipn.mx:8080/xmlui/handle/LITER_CIIDIROAX/83

[43] Avila V, Rivera D, Nava U, Czaja A, Garcia MC, Estrada JL, et al. Parasitismo Natural de *Spodoptera frugiperda* en maiz en la Comarca Lagunera Mexico. Southwestern Entoologist. 2023;**45**:195-202. DOI: 10.3958/059.048.0119

[44] Lopez MA, Martinez-Castillo AM, Garcia-Gutierrez C, Cortez-Mondaca E, Escobedo-Bonilla CM. Parasitoides y entomopatogenos asociados al gusano cogollero del maiz, *Spodoptera frugiperda* (Lepidoptera:Noctuidae) , en el Norte de Sinaloa. Southwestern Entomologist. 2018;**43**:867-881. DOI: 10.3958/059.043.0405

[45] Bahena-Juarez F, dc Lange E, Farnier K, Cortez-Mondaca E, Sanchez-Martinez R, Garcia-Perez F, et al. Parasitismo de gusano cogollero del maíz (*Spodoptera frugiperda* J. E. Smith) (Lepidoptera: Noctuidae) en el centro de Mexico. In: Proceedings: XXXIII Congreso Nacional de Control Biologico; Uruapan, Michoacan, Mexico. 2010. pp. 204-209

[46] Hye-Rin L, Tae-Ho A, Deok-Seo K, Bong-Kyu B. Four newly recorded species of the genus Meteorus Haliday (Hymenoptera: Braconidae: Euphorinae) from Korea. Journal of Asia-Pacific Biodiversity. 2017:545-547. DOI: 10.1016/j.japb.2017.10.001

[47] Fujie S, Shimuzu S, Tone K, Matsuo K, Meto K. Stars in subtropical Japan: a new gregarious *Meteorus* species (Hymenoptera, Braconidae, Euphorinae) constructs enigmatic star-shaped pendulous communal cocoons. Journal of Hymenoptera Research. 2021;**86**:19-45. DOI: 10.3897/jhr.86.71225

[48] Ruiz-Cancino E, Tejada L, Rivera-Cantu M. Contribución al conocimiento de los braconidos (Hymenoptera) de Tamaulipas y Nuevo Leon, México. Folia Entomologica Mexicana. 1990;**78**:199-208

[49] Braet Y, Van Achterberg C. New species of the genera Foenomorpha Szépligeti (Cenocoliinae) and *Chelonus* Panzer (Chelonina) (Hymenoptera: Braconidae), from French Guiana, Suriname, and Brazil. Zoologische mededeelingen. 2021;**75**:10-118

[50] Wharton R, Marsh P, Sharkey M. Manual of the new world genera of the Family Braconidae (Hymenoptera). USA: Madison; 1997

[51] Cave R. Parasitoides larvales y rurales de *Spodoptera frugiperda* (Smith) (Lepidoptera:*Noctuidae*) en Centro America con una clave para las especies encontradas en Honduras. Revista CEIBA. 1993;**34**:33-56

[52] Zhou J, Van Achterberg C, Tang P, Chen X. A new subgenus of *Chelonus* Panzer, 1806 (Braconidae: Cheloninae) from China. Zootaxa. 2022;**5115**:288-294. DOI: 10.11646/zootaxa.5115.2.8

[53] Jourdie V, Alvarez N, Turlings TCJ. Identification of seven species of hymenopteran parasitoids of *Spodoptera frugiperda,* using polymerase chain reaction amplification and restriction enzyme digestion. Agricultural and Forest Entomology. 2008;**10**:129-136. DOI: 10.1111/j.1461-9563.2008.00362.x

[54] Gonzalez-Maldonado MB, Garcia-Gutierrez C, Gonzalez-Hernandez A. Parasitismo y distribucion de *Campoletis sonorensis* Cameron (Hymenoptera: Ichneumonidae) and *Chelonus insulares* Cresson (Hymenoptera: Braconidae), parasitoides del gusano cogollero del maíz en Durango, Mexico. Vedalia. 2014;**15**:47-63

[55] Muesebeck CFW. A revision of the North American species of ichneumon-flies belonging to the genus *Meteorus* Haliday. Proceedings of the United States National Museum. 1923;**1923**(63):1-44. DOI: 10.5479/si.00963801.63-2470.1

[56] Aguirre FH, Sarmiento CE, Shaw SR. Taxonomic revision and morphometric analysis of *Meteorus* Haliday, 1835 (Hymenoptera: Braconidae: Meteorinae) from Colombia. Zootaxa. 2011;**2938**:1-68. DOI: 10.11646/zootaxa.2938.1.1

[57] Huddleston T. A revision of the western palearctic species of the genus *Meteorus* (Hymenoptera: Braconidae). Bulletin of the British Museum (Natural History); Entomology. 1980;**41**:1-58. Available from: http://www.biodiversitylibrary.org/page/2290999#page/11/mode/1up

[58] Aguirre H, Almeida LFV, Shaw SR, Sarmiento CE. An illustrated key to neotropical species of the genus *Meteorus* Haliday (Hymenoptera, Braconidae, Euphorinae). ZooKeys. 2015;**2015**(489):33-94. DOI: 3897/zookeys.489.9258

[59] Abe Y, Nishimura T, Maeto K. Causes of polymorphic melanism and its thermoregulatory function in a parasitoid wasp *Meteorus pulchricornis* (Hymenoptera: Braconidae). European

Journal of Entomology. 2013;**110**:627-632. DOI: 10.14411/eje.2013.085

[60] Folmer O, Black M, Hoeh W, Lutz R, Vrijenhoek R. DNA primers for amplification of mitochondrial cytochrome c oxidase subunit I from diverse metazoan invertebrates. Molecular Marine Biology and Biotechnology. 1994, 1994;**3**:294-299 mbari.org/Folmer_94MMBB

[61] Jourdie V, Eduardo V, Murillo H, Bento JM, Turlings TCJ, Alvarez N. Phylogeography of *Chelonus insularis* (Hymenoptera: Braconidae) and *Campoletis sonorensis* (Hymenoptera: Ichneumonidae), two primaries neotropical parasitoids of the fall armyworm (Lepidoptera: Noctuidae). Annals of the Entomological Society of America. 2010;**103**:742-749. DOI: 10.1603/AN09156

[62] Wernersson R, Pedersen AG. RevTrans: multiple alignment of coding DNA from aligned amino acid sequences. Nucleic Acids Research. 2003;**31**:3537-3539. DOI: 10.1093/nar/gkg609

[63] Thompson JD, Gibson TJ, Plewniak F, Jeanmougin F, Higgins DG. The CLUSTAL_X windows interface: Flexible strategies for multiple sequence alignment aided by quality analysis tools. Nucleic Acids Research. 1997;**25**:4876-4882. DOI: 10.1093/nar/25.24.4876

[64] Harrison CJ, Langdale JA. A step by step guide to phylogeny reconstruction. The Plant Journal. 2006;**45**:561-572. DOI: 10.1111/j.1365-313x.2005.02611.x

[65] Librado P, Rozas J. DnaSP v5: A software for comprehensive analysis of DNA polymorphism data. Bioinformatics. 2009;**25**:1451-1452

[66] Excoffier L, Lischer HL. Arlequin suite ver. 3.5: A new series of programs to perform population genetics analyses under Linux and Windows. Molecular Ecology Resources. 2010;**10**:564-567. DOI: 10.1111/j.1755-0998.2010.02847.x

[67] Harpending HC. Signature of ancient population growth in a low-resolution mitochondrial DNA mismatch distribution. Human Biology. 1994;**66**:591-600. Available from: https://www.jstor.org/stable/41465371

[68] Rogers AR, Harpending HC. Population growth makes waves in the distribution of pairwise genetic differences. Molecular Biology and Ecology. 1992;**9**:552-569

[69] Excoffier L. Patterns of DNA sequence diversity and genetic structure after a range expansion: Lessons from the infinite-island model. Molecular Ecology. 2004;**13**:853-864. DOI: 10.1046/j.1365-294X.2003.02004.x

[70] Rozas J. DNA sequence polymorphism analysis using DnaSP. In: Posada D, editor. Bioinformatics for DNA Sequence Analysis. Methods in Molecular Biology Series. Clifton, N.J: Humana Press; 2009. pp. 337-350

[71] Ronquist F, Huelsenbeck JP. MrBayes 3: Bayesian phylogenetic inference under mixed models. Bioinformatics. 2003;**19**:572-1574. DOI: 10.1093/bioinformatics/btg180

[72] Rannala B, Yang Z. Probability distribution of molecular evolutionary trees: A new method of phylogenetic inference. Journal of Molecular Evolution. 1996;**43**:304-311

[73] Huelsenbeck JP, Larget B, Miller RE, Ronquist F. Potential applications

and pitfalls of Bayesian inference of phylogeny. Systematic Biology. 2002;**51**:673-688. DOI: 10.1080/10635150290102366

[74] Avise JC. Phylogeography: The History and Formation of Species. Cambridge: Harvard University Press; 2000. p. 464. DOI: 10.2307/J.ctv1nzfgj7

Chapter 10

Productive and Economic Losses Caused by *Dichelops melacanthus* in Transgenic Bt Maize *Bacillus thuringiensis*

Marcos Arturo Ferreira Agüero, Felisa de Los Ríos Peixoto, Arsenio Benítez Sánchez, José Augusto Velásquez, Gustavo Daniel Vega-Brítez, Nelson David Lesmo Duarte and Matheus Francisco Acosta Resquín

Abstract

Transgenic maize expressing *Bacillus thuringiensis* (Bt) toxin produces a crystal (Cry) protein toxic to caterpillars that is non-toxic to stink bugs. The objectives of this study were to identify the number of plants attacked and not by *Dichelops melacanthus*, to evaluate foliar damage through the number of punctures, to evaluate plant height, weight of grain production by corn plants attacked and not and economic loss. The research was carried out on a commercial production agricultural farm in an area of 700 m^2. Eight areas were evaluated (10 m × 1.8 m) randomly distributed and in V6 physiological stage. Data were recorded, tabulated in Excel spreadsheet and statistically analysed by T Student test with 5% of significance for comparison of two independent groups. The results indicate that 80.7% of corn plants were attacked by *D. melacanthus* presenting punctures in their leaves. The average height for attacked plants was 41.2 ± 2.2 cm and 41.5 ± 3.3 cm for not attacked. A significant reduction in production of 23% was verified. The weight of grains of attacked plant was in average 3048 ± 319 g and 3956 ± 269 g in not-attacked plant, demonstrating that the damage caused by *D. melhacantus* reduces Bt corn productivity and loss of income of 98.93 US dollars per hectare.

Keywords: *Dichelops melacanthus*, Pentatomidae, phytophagous, Bt corn, insecticide

1. Introduction

Bt corn is a type of transgenic corn that produces a protein of bacterial origin. The Cry protein, naturally produced by *Bacillus thuringiensis*, is toxic to defoliating caterpillars or stem borers, but not toxic to stink bugs. Maize is cultivated practically throughout the Paraguayan territory with different production systems and technological levels used. Some changes in the corn production system have contributed to the increase in productivity: the direct sowing system, the use of hybrids with high productive potential, the increase in cultivated areas in the second harvest, after the soybean harvest, and the use of genetically modified hybrids such as Bt maize [1].

In recent years, the main problem faced by maize producers is the difficulty of protecting the maize crop from leaf damage caused by *Dichelops melacanthus* (Hemiptera: Pentatomidae) during the first weeks after maize emergence. At this stage, the seedling is quite susceptible to leaf punctures caused by Hemipteran stylets during feeding.

Studies to estimate the economic threshold level for *D. melacanthus* in maize were carried out by Bianco [2] who estimated two insects per five plants of corn, which is equivalent to 2 stinkbugs/1.25 lineal m under field conditions. In addition, Gassen [3, 4] and Cruz et al. [5] recommend control measures for stinkbugs in the corn crop when two insects/m^2 are found, which is equivalent to one bug/lineal m. Both levels are higher than that found by Duarte [6] with 0.8 stinkbugs/m^2 equivalent to 0.4 insects/linear m under controlled population conditions.

What is worrying is that both Bt technology and seed treatment with neurotoxic insecticides fail to provide protection to the maize plant in the initial vegetative phase, this forces producers to be vigilant and take early control actions so that economic damages do not render production unfeasible due to foliar damage and the increase in production costs. The question that arises is: What percentage of attack can a commercial corn crop suffer and how much can it reduce the production of corn grains in weight and economic income?

However, these changes can trigger new problems and require constant studies for proper management. The occurrence of new pests or the increase of others that attack the crop can be seen as a direct reflection of alterations in the productive systems. The appearance of *Dichelops melacanthus* (Hemiptera: Pentatomidae), commonly known as green-bellied bug, is currently found in Brazil, Argentina, Paraguay, Uruguay, Bolivia, Colombia, Peru and Venezuela [7]. Infestation and damage occur at the seedling stage as a result of adult migration from surrounding crop debris or other plants within the field [8–10].

A transformation of great magnitude, such as the one experienced in our agriculture, should respond to a reasoned process, supported by knowledge and adequate technological management of each component of the production system. One of the fundamental pillars on which the cultivation of corn rests and that, therefore, directly influences the yields achieved, is pest control.

Knowledge of the population dynamics of these pests is important in the management of these organisms since through it the incidence is estimated and the management of the insect can be planned, which is essential when determining a control strategy that avoids the increase in the existing gap between potential returns and real returns [11]. Damage to maize from seedlings causes brown spots, leaf discolouration and twisting, reduced yield [12] or plant death [10, 13].

The realization of this work is based on the current concern about the high incidence of *D. melacanthus* in the corn crop, even with seed treatment and a series of insecticide applications during the crop cycle. This research tries to obtain real data from the field that can contribute to verify the level of damage caused by the attack of bugs in corn production and how this can affect the producer economically.

In this research, the objectives have been to evaluate the foliar damage caused by the attack of the green-bellied bug *D. melacanthus* on transgenic Bt maize, grain production, to identify the number of attacked and non-attacked plants. Quantify the number of punctures per plant, verify the height of the plant in the phenological state V6 of maize and determine the weight of grains of the attacked and unattacked plants to compare the yield.

2. Materials and methods

The study was carried out in the 2016 harvest at the Gredos Agricultural Farm, of the Martin and Martin Group, located 12 km from the city of Pedro Juan Caballero, in the Vice President Sánchez neighbourhood, on route 5 "General Bernardino Caballero", Department of Amambay, georeferenced at 22°36′17″S 55°48′46″W. The property is located at approximately 640 m above sea level, the topography is flat and the area is used for extensive crops under a direct sowing system in rotation with soybeans, second-harvest corn, wheat and chia.

The region is characterized by having a transitional climate between a Mediterranean type and a frankly humid climate with an average annual temperature of 22°C and an average annual rainfall of between 1000 and 1200 mm per year, with the rainiest months being December and January, the less rainy months June, July and August. The soil of the place corresponds to an alfisol (Soil Taxonomy), with a clayey loam texture and shows a prolonged agricultural use of more than twenty-seven years of sustained production, with twelve years of production of cereals and other grains under the direct sowing system.

The study area was made up of eight areas randomly distributed over a 700 m^2 surface, the dimensions of the areas being 20 m wide and 35 m long with five rows of corn separated 0.45 m by 10 m long, with an average population of 108.2 plants. For comparison purposes, they were classified into plants attacked and not attacked by *D. melacanthus*.

The criteria adopted to classify the attacked plants consisted of the presence of three to four holes in the corn leaf blade that generally appear in three to four rows and are characteristic damages that form on the leaves after the puncture-puncture performed by the stylets from bed bugs. Meanwhile, for the plants not attacked, those maize plants that did not show the characteristic damage described were considered.

The number of Bt maize plants attacked and not attacked by *D. melacanthus* and the number of punctures per plant, height of the plant in the V6 phenological stage and the weight of grains of the attacked and non-attacked Bt maize plants have been quantified by *D. melacanthus*. The estimation of attacked and unattacked plants, number of punctures and height of the plant, was carried out 33 days after sowing, and the grain weight of the attacked and unattacked plants was carried out after harvest.

The calculation of the percentage of plants attacked and/or attacked was carried out through the analysis of relative frequency (hi) = fi/N, where fi corresponds to the number of times that a piece of data is repeated within the set and N, the total number of data from the set.

The unattacked corn plants were marked with a red tape, in all the areas attacked plants were found, that is, approximately 85% of the population was attacked by green-bellied bugs, even after making two applications of the insecticide Imidacloprid to control the pest, even stinkbugs continued to appear in the areas, so a third application of the product was made. In total, 866 Bt maize plants were evaluated.

In the demarcated areas, the number of punctures caused by the green-bellied bug was quantified, which was recorded by direct observation of the continuous perforations in each leaf and noted on a spreadsheet. The height of the plant was measured with a tape measure in the phenological stage V6 of maize, at that stage the plants presented a high attack by stinkbugs.

Once the maize reached commercial maturity, the ears were harvested manually. For this, the ears of corn were pulled out and placed in bags, which received an identification. Subsequently, the ears were exposed to the sun for 12 hours to reduce their humidity, and then the shelling and cleaning were carried out manually to eliminate impurities. The production was packed in the bags and identified according to their respective areas. Finally, the corn grains were weighed on a Profield brand four-digit precision scale.

The estimation of the corn yield was made by multiplying the average weight of the corn grains by the number of existing plants in the evaluated area, later, by the simple rule of three, the value of the corn yield per hectare was estimated. The calculation of the loss of income was made by multiplying the average weight loss of grams of corn per hectare by the commercial sale value of the product.

The data were arranged into two groups, a population of plants attacked by *D. melacanthus* and another population of plants not attacked by *D. melacanthus*. To verify the normal distribution, the data referring to the number of attacked and/or attacked plants as well as the height of the plants were subjected to the Shapiro Wilk test at 5% significance, the analysis showed that the data were homogeneous and met the requirements for the distribution. Use of parametric statistics, by and analyzed, using the Student's T-test at 5% significance for comparison of two independent groups and verification of the existence of significant differences between treatments. The statistical program used was BioStat 5.3.

3. Results

It has been verified that of the 866 Bt maize plants evaluated in the vegetative physiological stage V6, 699 (80.7%) of the plants were attacked by the *D. melacanthus* stinkbug. The maize leaves presented evident damage by the presence of punctures or holes produced by the feeding of bed bugs with an average number of 6.1 punctures per plant. On the other hand, 167 maize plants (19.9%) were not attacked or did not present the characteristic symptoms described for *D. melacanthus* (**Table 1**).

The average number of *D. melacanthus* plants attacked was 87.4 ± 7.8 and 20.8 ± 2.79 not attacked, from an average population of 108.2 plants, considering the eight Bt maize areas evaluated in which there were between 100 and 117 plants.

The high value of maize plants attacked is due to the fact that the Cry protein, expressed by the Bt event, the seed treatment and the foliar application with neurotoxic insecticides were unable to provide protection to the maize plant from the

Treatments	No. of plants (n = 866)	Average number of plants attacked and not attacked	Height (cm)
Attacked plants	699	87.4 ± 7.8 a	41.2 ± 2.2 a
Plants not attacked	167	20.8 ± 2.7 b	41.5 ± 3.3 a

Average values followed by different letters in columns differ from each other by Student's T-test at 5% significance.

Table 1.
Number of Bt maize plants and height of plants (cm) attacked by D. melacanthus *and not attacked.*

attacks of *D. melacanthus*. On the one hand, the protein crystals of *B. thuringiensis* are non-toxic to pentatomidae bugs and on the other hand, the insecticide imidacloprid applied *via* seed and foliar was inefficient in controlling the population of pest hemipterans.

In relation to the variable height of the Bt maize plants, attacked or not by *D. melacanthus*, no statistically significant difference ($p > 0.05$) was verified between the populations. An average height of 41.5 cm was recorded for unattacked plants and an average height of 41.2 cm for attacked plants (**Table 1**).

These results reveal that the height growth of maize plants is not affected, however, the leaf area is compromised by the feeding punctures of the stink bugs, which in turn affect the photosynthetic capacity of the plant and the translocation of photoassimilates to the drainage region for the formation of corn kernels, which reduces their weight.

It has been verified that the average weight of grains per Bt maize plant not attacked by *D. melacanthus* was 3956 ± 269 g and for the attacked plants it was 3048 ± 199 g with significant statistical differences between them ($p < 0.05$). It has also been estimated that the average yield per hectare of the unattacked Bt maize plants was 2197.77 kg/ha and those attacked, 1693.33 kg/ha, thus the loss in maize production was 504.44 kg /ha due to damage caused by *D. melacanthus*. In this sense, considering the average value of corn in the corresponding harvest, the loss of income was 98.93 dollars per hectare (**Figure 1**).

4. Discussion

According to the data referring to plants attacked and not, a high incidence of the green-belly bug has been verified in the Bt maize crop itself with applications of Imidacloprid insecticides (200 ml/ha). Probably the high occurrence of *D. melacanthus* is due to the resistance to the active principles applied during the development of

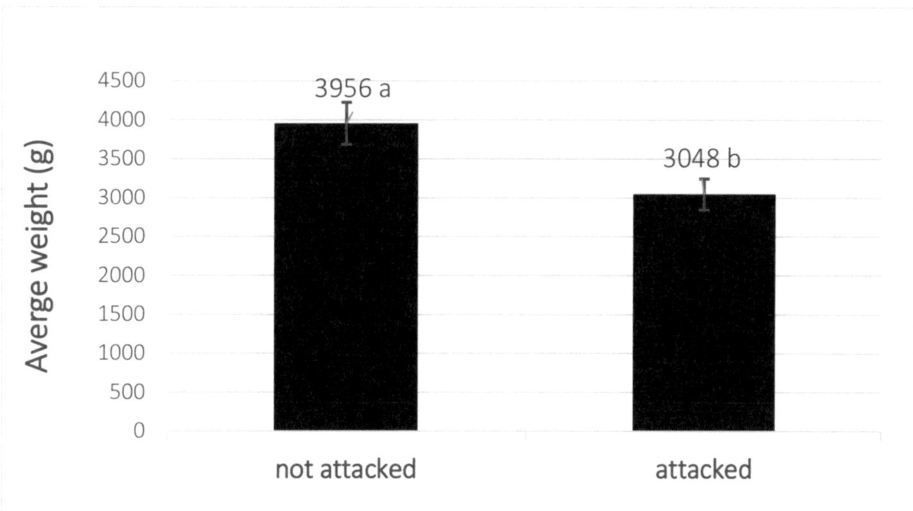

Figure 1.
Average kernel weight ± SD (g) per Bt maize plant not attacked and attacked by Dichelops melacanthus. *Average values followed by different letters differ from each other by Student's T-test at 5% significance.*

maize. Another aggravating factor is that soybeans grown before corn also host a large number of bugs and the same chemicals are used to control them, which generates selection pressure for resistant biotypes.

No significant differences were observed in the height decrease of attacked and unattacked maize plants. This is compatible with those obtained by Crosariol Netto et al. [12], who observed that the transgenic plants did not show significant differences in relation to the height and respond differently to conventional non-Bt hybrids than if they can reduce plant height due to stink bug damage.

A high incidence and damage by *D. melacanthus* has also been verified in the V6 vegetative stage, which is consistent with the work of Copatti and Oliveira [14]. In their study, they show a high potential for damage to maize, which occurs in the initial state of development between V2 and V8. In other similar works, the reduction in plant height was not significant between conventional and transgenic hybrids, but rather it is a behaviour between varieties [15]. In another similar work, it is highlighted that the height of maize plants is not affected in any of the population and infestation of *D. melacanthus* [16].

As verified in this study, significant damage in maize is observed especially if the infestation occurs in the physiological stages V1 and V3, negatively affecting crop production [6], confirming a decrease in maize yield. (kg/ha) by increasing the number of green-bellied bugs per square meter, as observed in this research. In other similar works, the damage to the productivity components themselves is observed with field infestations of 2 and 4 bugs per m^2 [17]. Despite the reduction in plant height in conventional hybrids, no effects on productivity are observed, thus being the main factor in yield reduction due to a high infestation of *D. melacanthus* [12].

The same authors did not observe a relationship between the average weight of 100 corn kernels and the density of *D. melacanthus*. However, a negative relationship was revealed between the average weight of the spikes and the population densities of *D. melacanthus*. However, the grain yield decreases with the increase in the levels of infestation of stink bugs, evidencing that the increase in the population density of the insect reduces the weight and yield of grains in the corn crop. Bridi et al. [16] highlight that the reduction in productivity is 7.1% for each *D. melacanthus* added in 1 m^2, in a range of 0–4 stinkbugs per m^2.

Portela et al. [18] obtained similar results, where they verified that the green-bellied bug causes a greater intensity of reduction in the weight of the maize grain when compared to the brown bug *Euschistus heros*, evidencing that the first species potentially cause greater damage to this crop than the second.

Bridi et al. [16] highlight the reduction of grains of up to 3.96 grains per row, which represents a decrease of 12.3% for every 4 stinkbugs per m^2 compared to the absence of it. The length of the spike is also affected by the infestation of 3.16 stinkbugs, which causes a reduction of about 12% in relation to the size of the spike without the presence of the insect. In other similar works such as that of Cruz et al. [19], it is described that without the presence of the insect, the yield was 8048.43 kg/ha, while, in the presence of the insect, the grain yield was 6352.21 kg/ha, a difference of 21.07% or 1696.22 kg/ha.

According to Duarte et al. [6], it is possible to estimate that the level of economic damage for *D. melacanthus* in the corn crop is 8 bugs/m^2, population density above which pest control is economically justified. These results disagree with those obtained by Bridi et al. [16] where they indicate that between 1 and 4 bugs per m^2 significantly affects corn yield.

The insecticide commonly used to control bugs in corn is Imidacloprid. Research carried out by Chiesa et al. [8] show that this product can reduce the population density of the bug between 23.2 and 61.8% also that seed treatment with this insecticide does not efficiently protect against the attack of the bug *D. melacanthus*.

Albuquerque et al. [20] show that Imidacloprid may not be efficient when applied 8 days after the emergence of corn plants. Also, pre-emergence insecticide sprays have little effect on *D. melacanthus*, while post-emergence applications can achieve up to 80% control [21].

Another reason that may favour the high incidence of *D. melacanthus* is high temperature. The *D. melacanthus* bug performs better in high-temperature conditions (up to 31 ± 1°C) while constant temperatures of 19°C harm it. It has also been shown that Bt events such as transgenic soybeans do not affect their biology [22].

In several studies it has been shown that the treatment of corn seeds does not efficiently protect against the attack of *D. melacanthus*, also that the series of applications of chemical products in the vegetative and reproductive stages does not efficiently reduce the population density of the green-bellied bug. According to Modolon et al. [23], control plants without chemical seed treatments can present up to 100% of the plants attacked by *D. melacanthus*. On the other hand, Brustolin et al. [21] recorded up to 60% of plants attacked by *D. melacanthus* without seed treatments, while with treatments, the number of attacked plants can be reduced to 24%.

5. Conclusions

Most of the Bt maize plants (80.7%) were attacked by *D. melacanthus*. Which reduced productivity and economic income.

Dichelops melacanthus caused an average of 6.1 punctures per Bt maize plants, affecting grain yield by 23% and causing a loss of income of 98.93 US dollars per hectare.

Average kernel weight of Bt maize plants was reduced due to damage caused by *D. melacanthus*. However, the reduction in the average height of Bt maize plants was not significant.

In future research, it is recommended to evaluate the effect of the attack of the bugs in the reduction of leaf area and photosynthetic capacity that reduces maize production. Other control methods for *D. melacanthus* should also be evaluated.

Acknowledgements

National Council of Science and Technology (CONACYT, Paraguay) and National Program of Incentive for Researchers (PRONII), PROCIENCIA.

Conflict of interest

The author declares no conflict of interest.

Additional information

This work is the English translation of the article: Ferreira-Aguero MA, Benítez-Sánchez A, Velásquez JA, Vega-Britez GD, Lesmo-Duarte ND, Acosta-Resquín MF. Daños causados por chinche barriga verde Dichelops melacanthus en maíz transgénico *Bacillus thuringiensis* (Bt) [Internet]. Intropica. Universidad del Magdalena; 2020. pp. 66–71. Available from: http://dx.doi.org/10.21676/23897864.3938.

Author details

Marcos Arturo Ferreira Agüero*, Felisa de Los Ríos Peixoto, Arsenio Benítez Sánchez, José Augusto Velásquez, Gustavo Daniel Vega-Brítez, Nelson David Lesmo Duarte and Matheus Francisco Acosta Resquín
Faculty of Agrarian Sciences, National University of Asunción, Pedro Juan Caballero, Amambay, Paraguay

*Address all correspondence to: marcos.ferreira@agr.una.py

References

[1] Ortigoza G, López TCA, González VJD. Cultivo de maíz. Guía técnica. San Lorenzo, Paraguay: JICA/FCA-UNA; 2019. p. 48. Available from: https://www.jica.go.jp/paraguay/espanol/office/others/c8h0vm0000ad5gke-att/gt_04.pdf [Consultado: 12 de enero 2020]

[2] Bianco R. Nível de dano e período crítico do milho ao ataque do percevejo barriga verde (Dichelops melacanthus). In: Congresso Nacional de Milho e Sorgo, 25. Simpósio Brasileiro Sobre a Lagarta do Cartucho *Spodoptera frugiperda*, 1. Cuiabá. Resumos, Sete Lagoas: ABMS: Embrapa Milho e Sorgo: Empaer; 2004. p. 172

[3] Gassen DN. Manejo de pragas associadas à cultura do milho. Passo Fundo: Aldeia Norte; 1996. p. 127

[4] Gassen DN. Pragas associadas à cultura do milho. Passo Fundo: Aldeia Norte; 1994. p. 92

[5] Cruz I, Viana PA, Waquil JM. Manejo das pragas iniciais de milho mediante tratamento de sementes com inseticidas sistêmicos. Sete Lagoas: Embrapa-CNPMS; 1999. p. 39

[6] Duarte MM, Ávila CJ, Santos V. Danos e nível de danoeconômico do percevejo barriga verde na cultura do milho. Revista Brasileira de Milho e Sorgo. 2015;**14**(3):291-299. DOI: 10.37856/bja.v84i2.1494

[7] Dellapé G, Rider DA, Dellapé DA. Notes on distributions for Argentine an Pentatomidae (Heteroptera: Pentatomoidea), with new records in the country. Revista Brasileira de Entomologia. 2015;**59**(3):169-176. DOI: 10.1016/j.rbe.2015.06.001

[8] Chiesa ACM, Sismeiro MNS, Pasini A, Roggia S. Tratamento de sementes para manejo do percevejo barriga verde na cultura de soja e milho em sucessão. Pesquisa Agropecuária Brasileira. 2016;**51**(4):301-308. DOI: 10.1590/S0100-204X2016000400002

[9] Smaniotto LF, Panizzi AR. Interactions of selected species of stink bugs (Hemiptera: Heteroptera: Pentatomidae) from leguminous crops with plants in the neotropics. Florida Entomologist. 2015;**98**(1):7-17. Available from: https://www.jstor.org/stable/24364166?seq=1 [Consultado: 9 de enero 2020]

[10] Torres ABA, de Oliveira NC, de Oliveira-Neto AM, Guerreiro JC. Injúrias causadas pelo ataque dos percevejos marrom e barriga verde durante o desenvolvimento inicial do milho. Journal of Agronomic Sciences. 2013;**2**(2):169-177

[11] Vázquez-Porras I, González-Gaona O, Espinosa-Vásquez G, Terán-Vargas AP, Azuara-Domínguez A. Dinámica poblacional y asociación de insectos plaga y benéficos a las fases fenológicas del cultivo de soya en el sur de Tamaulipas. Entomología Mexicana. 2016;**3**:305-310

[12] Crosariol Netto J, Michelotto MD, Grigolli JFJ, Galli JA, Pirotta MZ, Busoli AC. Danos ocasionados por Dichelopsmelacanthus (Heteroptera:Pentatomidae) em híbridos convencionais e transgênicos de milho. Bioscience Journal. 2015;**31**(4):1092-1101

[13] Roza-Gomes MF, Salvadori JR, Pereira PRV, da Panizzi AR. Injúrias de quatro espécies de percevejos pentatomídeos em plântulas de milho.

Ciencia Rural. 2011;**41**:1115-1119. DOI: 10.1590/S0103-84782011005000081

[14] Copatti JF, Oliveira NC. Danos iniciais causados pelos percevejos *Dichelops melacanthus* e *Euschistus heros* (Hemiptera: Pentatomidae) em plantas de milho. Campo Digital. 2011;**6**(1):1-8. Available from: http://revista2.grupointegrado.br/revista/index.php/campodigital/article/view/1126 [Consultado: 6 de enero de 2020]

[15] Michelotto MD, Crosariol Netto J, Freitas R, Duarte A, Busoli A. Milho transgênico (Bt): efeito sobre pragas alvo e não alvo. Nucleus. 2013;**10**(3):67-82. DOI: 10.3738/nucleus.v0i0.903

[16] Bridi M, Kawakami J, Hirose E. Danos do percevejo Dichelops melacanthus (Dallas, 1851) (Heteroptera: Pentatomidae) na cultura do milho. Magistra. 2017;**28**(3/4):301-307. Available from: https://magistraonline.ufrb.edu.br/index.php/magistra/article/view/283 [Consultado: 2 de enero de 2020]

[17] Bridi M. Danos de percevejos pentatomídeos (Heteroptera: Pentatomidae) nas culturas da soja e do milho na região centro-sul do Paraná [Doctoraldissertation, Tese de mestrado]. Guarapuava. Available from: http://unicentroagronomia.com/destino_arquivo/dissertacao_marcelo_bridi.pdf: Universidade Estadual do Centro Oeste; 2012 [Consultado: 2 de enero de 2020]

[18] Portela ACV, Santos V, Salvador DJ, Ávila CJ. Danos causadospelos percevejosEuschistusheros (Fabricius) e Dichelopsmelacanthus(Dallas) (Heteroptera: Pentatomidae) em trigo e milho. In: In: XXI Congresso Brasileiro de Entomologia, Recife PE, Anais... Recife: UniversidadeFederal Rural de Pernambuco, CD-ROM de Resumos; 2006

[19] Cruz I, Bianco R, Redoam A. Riesgo potencial de pérdidas en el maíz causadas por Dichelops melacanthus (Dallas) (Hemiptera: Pentatomidae) en Brasil. EmbrapaMilho y Sorgo-Artigo en periódico indexado (ALICE). 2016. DOI: 10.18512/1980-6477/rbms.v15n3p386-397 [Consultado: 6 de enero de 2020]

[20] Albuquerque FA, Borges LM, Iacono TO, Crubelati NCS, Singer AC. Eficiência de inseticidas aplicados em tratamento de sementes e em pulverização, no controle de pragasiniciais do milho. Revista Brasileira de Milho e Sorgo. 2006;**5**(1):15-25. DOI: 10.18512/1980-6477/rbms.v5n01p%25p

[21] Brustolin C, Bianco R, Neves PMOJ. Inseticidas em pré e pós-emergência do milho (*Zea mays* L.), associados aotratamento de sementes, sobre Dichelopsmelacanthus (Dallas) (Hemiptera: Pentatomidae). Revista Brasileira de Milho e Sorgo. 2011;**10**(3):215-223. DOI: 10.18512/1980-6477/rbms.v10n3p215-223 [Consultado: 10 de enero de 2020]

[22] Bortolotto OC, Mikamir AY, Bueno AF, Silva GV, de Queiroz AP. Aspectos biológicos de Dichelopsmelacanthus em três temperaturas, alimentados com grãos imaturos de milho 2B688Hx e 2B688. Ciencia Rural. 2016;**46**:254-259. DOI: 10.1590/0103-8478cr20150542

[23] Modolon TA, Pietrowski V, Alves LFA, Guimarães ATB. Desenvolvimento inicial do milho tratado com o preparodo homeopático Nuxvomica e submetido ao percevejo barriga-verde Dichelops melacanthus Dallas (Heteroptera: Pentatomidae). Revista Brasileira de Agroecologia. 2016;**11**(2):85-93